Durability of Concrete

Modern Concrete Technology Series

A series of books presenting the state-of-the-art in concrete technology

Series Editors
Arnon Bentur
Faculty of Civil and Environmental Engineering Technion-Israel Institute of Technology

Sidney Mindess
Department of Civil Engineering University of British Columbia

Durability of Concrete: Design and Construction
Mark Alexander, Arnon Bentur, Sidney Mindess   Hb: 978-1-138-74674-9

Sustainability of Concrete
Pierre-Claude Aïtcin, Sidney Mindess   Hb: 978-1-138-07568-9

Binders for Durable and Sustainable Concrete
P. -C. Aïtcin   Hb: 978-0-415-38588-6

Fibre Reinforced Cementitious Composites 2nd Edition
A. Bentur & S. Mindess   Hb: 978-0-415-25048-1

Diffusion of Chloride in Concrete
E. Poulsen & L. Mejlbro   Hb: 978-0-419-25300-6

Pore Structure of Cement-Based Materials: Testing, Interpretation, and Requirements
K. K. Aligizaki   Hb: 978-0-419-22800-4

Aggregates in Concrete
M. G. Alexander & S. Mindess   Hb: 978-0-415-25839-5

Fundamentals of Durable Reinforced Concrete
M. G. Richardson   Hb: 978-0-419-23780-8

Sulfate Attack on Concrete
J. Skalny, J. Marchand & I. Odler   Hb: 978-0-419-24550-6

Special Inorganic Cements
I. Odler   Hb: 978-0-419-22790-8

Concrete Mixture Proportioning
F. de Larrard   Hb: 978-0-419-23500-2

High Performance Concrete
P. -C. Aïtcin   Hb: 978-0-419-19270-1

Durability of Concrete
Design and Construction

**Mark Alexander, Arnon Bentur,
and Sidney Mindess**

CRC Press
Taylor & Francis Group
Boca Raton London New York

CRC Press is an imprint of the
Taylor & Francis Group, an **informa** business

CRC Press
Taylor & Francis Group
6000 Broken Sound Parkway NW, Suite 300
Boca Raton, FL 33487-2742

International Standard Book Number-13: 978-1-1387-4674-9 (Paperback)
978-1-4822-3725-2 (Hardback)

Library of Congress Cataloging-in-Publication Data

Names: Alexander, Mark G., editor. | Bentur, Arnon, editor. | Mindess, Sidney, editor.
Title: Durability of concrete : design and construction / [edited] by Mark Alexander, Arnon Bentur and Sidney Mindess.
Other titles: Durability of concrete (CRC Press)
Description: Boca Raton : CRC Press, [2017] | Series: Modern concrete technology series | Includes bibliographical references and indexes.
Identifiers: LCCN 2017002663| ISBN 9781482237252 (hardback : alk. paper) | ISBN 9781138746749 (pbk. : alk. paper) | ISBN 9781482237269 (ebook : alk. paper)
Subjects: LCSH: Concrete--Deterioration. | Concrete--Service life.
Classification: LCC TA440 .D81915 2017 | DDC 624.1/834--dc23
LC record available at https://lccn.loc.gov/2017002663

Visit the Taylor & Francis Web site at
http://www.taylorandfrancis.com

and the CRC Press Web site at
http://www.crcpress.com

To our long-suffering wives

Lyn
Esty
Joanne

Contents

Preface

Concrete has been by far the most widely used construction material in the world for most of the last 100 years, and this is unlikely to change in the foreseeable future. However, concrete science and technology have changed considerably over this time, and the rate of this change has accelerated greatly over the past several decades. There are a number of reasons for this change, including

- A much more detailed understanding of the hydration reactions and the microstructure (and now the nanostructure) of the hardened cement paste and their relationships with engineering properties, which has permitted change based on science rather than on "trial-and-error" experimentation
- The development of much more efficient superplasticizers
- The introduction of fiber-reinforced concrete
- Increasing demands for specialized concretes which require enhanced concrete properties, that is, self-compacting concretes, concretes that can be pumped to heights of over 600 m, concrete strengths of over 200 MPa, increased rates of strength gain, and so on
- Growing demands by infrastructure and building owners for "maintenance-free" concrete structures that have an assured service life

As a result, the concretes that we use today are very different from those that the three authors of this book first worked with about 40 years ago. However, the great progress achieved in cement and concrete technology over this time has come with a price: today's concretes are much more sensitive to exactly how they are proportioned, mixed, placed, and cured. Despite our increased knowledge, the number of material failures in concrete construction does not appear to have decreased over this time. One of the contributing factors to this state of affairs is the increasing demand for more sustainable concrete, which in turn has led to the use of higher replacement levels of the Portland cement, and a much wider array of cement extenders (fillers and/or supplementary cementing materials). Unfortunately, the long-term behavior of these more complex concretes is often uncertain.

The most effective ways of making concrete more sustainable are by substantially increasing not only the amount of cement replacement, but also the service life of the concrete, that is, by increasing its durability. Therefore, this book is aimed mainly at the practicing engineer who increasingly is faced with durability issues in concrete design and construction. Currently, emphasis is shifting inexorably to "design for durability," and there is growing pressure from owners and infrastructure agencies for designers and constructors to actually deliver durable concrete! Engineers need tools to handle these challenges. This book provides an up-to-date survey of durability issues, including durability design and specifications, and how to achieve durability in actual concrete construction. The aim is to gather together aspects of concrete durability that presently are dispersed over diverse texts and sources, into one useful book. More specifically, the objectives are the following:

- To place durability into the context of modern concrete design and construction
- To summarize in understandable terms how and why concrete deteriorates
- To provide information on modern concrete materials and how they can be used to produce durable concrete
- To give a background to current philosophies around durability design and specifications, with reference to both European and North American practice
- To provide a clear understanding of how to achieve durability in actual construction

We hope that this book will be a valuable resource not only for designers and constructors of concrete structures, but also for students in graduate-level programs in universities and colleges.

Mark Alexander
Arnon Bentur
Sidney Mindess

Acknowledgments

Mark Alexander would like to acknowledge the invaluable assistance from various people in compiling Chapters 4, 6, and 10:

- Postgraduate students at the University of Cape Town, in particular, Dr. M. Kiliswa, Dr. R. Muigai, and Ms. G. Nganga
- Dr. M. Otieno from the University of the Witwatersrand
- Mr. G. Evans of PPC Cement Ltd.

Authors

Mark Alexander is emeritus professor of civil engineering in the University of Cape Town, South Africa, and former president of RILEM, the International Union of Laboratories and Experts in Construction Materials, Systems, and Structures.

Arnon Bentur is a professor emeritus and former vice president and director general of the Technion, Israel, and is former president of RILEM.

Sidney Mindess is a professor emeritus and former associate vice president at the University of British Columbia, Canada.

Chapter 1

Introduction

Concrete is by far the most common and widely used construction material. It is versatile, adaptable, economical, and, if properly made, durable. Its constituents (primarily Portland cement, aggregates, and water) are widely available. Indeed, next to water, it is the most widely used material in the world (Table 1.1), with perhaps 10 billion cubic meters produced annually (Felice, 2013), and this is unlikely to change much in the foreseeable future. Indeed, reinforced concrete accounts for more than half of all of the manufactured materials and products produced worldwide (Scrivener, 2014). However, a great deal of energy is required for the pyroprocessing of Portland cement. Of greater concern, a large amount of CO_2 is liberated in the production of Portland cement; depending on the plant efficiency, the quality of the raw materials, and the proximity of the cement plant to the raw materials, approximately 0.8 tons of CO_2 are released into the atmosphere per ton of Portland cement produced. Because of the huge volumes of concrete produced, this amounts to about 5%–8% of the world's total CO_2 emissions.

Though these numbers may seem alarming, it must be emphasized that, in terms of its carbon footprint, concrete is in fact a low-impact material. Concrete is a remarkably good and efficient construction material; if it was to be replaced by any other material, this would have a bigger carbon footprint. Concrete itself has a large carbon footprint simply because of the huge quantities of the material that are produced. Relative figures for the embodied energy and CO_2 emissions for some common building materials are shown in Table 1.2 (taken from Scrivener [2014], using data from Hammond and Jones [2008]). (*Note*: These figures are per kilogram; they do not take into account differences in specific strength of the various materials.)

It must also be noted that concrete is an inherently durable material. Of course, we no longer subscribe to the view that "*a reinforced concrete structure will be safe for all time, since its strength increases with age, the concrete growing harder and the bond with the steel becoming stronger*" or that "*its life is measured by ages rather than years*" (Thompson, 1909, p. 12). However, if concrete is correctly designed to operate in the environment to which it will be exposed, and if it is properly batched, placed,

1

Table 1.1 Annual worldwide production of materials, 2014 (tonnes)

Portland cement	4.3 billion
Concrete (estimated)	~23 billion
Coal	~7.8 billion
Steel	1.66 billion
Wood	~2.2 billion
Crude oil	~4.2 billion
Wheat	709 million
Salt	270 million
Sugar	173 million
Gold	2860

(The total amount of gold produced throughout all of world history would occupy about a 21-m cube!)

Table 1.2 Embodied energy and associated CO_2 emissions for some common construction materials

Material	Embodied energy (MJ/kg)	CO_2 (kg CO_2/kg)
Normal concrete	0.95	0.13
Fired clay bricks	3.00	0.22
Glass	15.00	0.85
Wood (plain timber)	8.5	0.46
Wood (multilayer board)	15	0.81
Steel (from ore)	35	2.8

Source: Adapted from Scrivener, K. L., 2014, Options for the future of cement, *The Indian Concrete Journal*, 88(7), 11–21; From data of Hammond, G. P. and Jones, C. I., 2008, Embodied energy and carbon in construction materials, *Proceedings of the Institution of Civil Engineers—Energy*, 161(2), 87–98.

and cured, it can perform its design function for a very long time. The Pantheon (Figure 1.1) and the Pont du Gard (Figure 1.2) are testaments to the potential longevity of concrete structures. In more recent times, the Ingalls Building, Cincinnati (Figure 1.3) was the world's first reinforced concrete skyscraper, standing 16 stories (64 m) high. The 6-story Hotel Europe, Vancouver (Figure 1.4), the first reinforced concrete structure in Canada, was built in 1908–1909. The Hoover Dam on the Colorado River (Figure 1.5), constructed in 1931–1936, was the largest concrete structure built to that time, using some 2,480,000 m^3 of concrete. These structures, and many others also still in use, are further examples of the potential extended service life of concrete.

The deterioration of concrete may be due to chemical, physical, or mechanical factors; these factors may be internal or external to the concrete structure. We will here be concerned primarily with the chemical and physical factors, which are the ones most influenced by the environment to which

Figure 1.1 The Pantheon, Rome, ~126 AD. (Courtesy of Katherine Mindess.)

the concrete is exposed. It must be emphasized that the various deterioration factors rarely occur in isolation; most commonly, several of these will act upon the concrete at the same time, and this will exacerbate the concrete deterioration. Unfortunately, there has been relatively little research dealing with the simultaneous action of two or more deterioration mechanisms, but it is clear that their effects are not simply additive. Chapter 4 provides more detail on important concrete deterioration mechanisms.

The mechanisms involved in the environmental deterioration of cement and concrete are by now well known, and there are a number of books

Figure 1.2 The Pont du Gard at Nîmes, France, built ~40–60 AD.

(and countless papers) that describe these mechanisms in detail. Why then still another concrete durability book? The focus here is not so much on the details of the chemical and physical processes that lead to concrete deterioration, though many of these will inevitably be reviewed briefly. Rather, the focus of this book is on the interrelationships among durability, sustainability, and economics.

"Concretes" (materials made using aggregates, a binder material, and water) go back many millennia, at least 6,000–12,000 years, when lime mortars were used in Crete, Cyprus, Greece, and other parts of the Middle East. Somewhat later, gypsum mortars were used by the Egyptians in the construction of the Pyramid of Cheops (about 4700 years ago). The Greeks and Romans made extensive use of lime–pozzolan mortars; the Pantheon and the Pont du Gard at Nîmes, amongst many others, attest to their mastery of concrete construction. However, "modern" concretes only go back to the early nineteenth century, when Joseph Aspdin, a Leeds builder, patented a material that he called "Portland cement." This was subsequently modified by his son William to produce a cement very close to that in use today.

When the authors of this book were first introduced to concrete, in the 1960s and 1970s, it was still a relatively simple material, at least in terms of its constituents. Supplementary cementing materials (SCMs) were still not in common use, and there were only a few chemical admixtures: air entraining agents, water reducers, retarders, and accelerators. Concrete consisted of "pure" Portland cement, aggregates, and water, plus perhaps one or two of the chemical admixtures mentioned. In most cases, not much

Figure 1.3 The Ingalls Building, Cincinnati, the first "skyscraper" at 16 stories, 1903. Its construction accounted for about 0.5% of all of the cement used in the United States in 1902–1903! (Courtesy of Collection of the Public Library of Cincinnati and Hamilton County.)

was demanded of such concretes: a slump of 75–100 mm and a compressive strength in the range of 15–30 MPa. *Durability* was only a secondary consideration in the mix design procedures of those days, and the concept of *sustainability* had not yet penetrated the minds of civil engineers and architects.

How much has changed since then! Strength *per se* is no longer much of an issue. With the widespread availability of SCMs, superplasticizers, and various other admixtures, it is not difficult to achieve strengths of 100 MPa or more. Paradoxically, however, these advances in concrete science and technology, along with minor changes in the composition of the Portland cement, have led to concretes that are less "forgiving" than those in earlier

Figure 1.4 The Hotel Europe, Vancouver, the first reinforced concrete structure in Canada, 1908–1909. (Courtesy of Katherine Mindess.)

days. That is, today's concretes are much more sensitive to their detailed chemistry, and how they are mixed, placed, and cured. As a consequence, they can be more prone to durability problems; this is particularly the case for high-performance concretes made with several SCMs, superplasticizers, and low water/binder ratios (see Chapter 3 for a more detailed discussion).

There has been a great deal of focus in recent years on how to make concretes more sustainable, particularly in terms of their carbon footprint.

Figure 1.5 The Hoover Dam on the Colorado River, then the largest concrete structure in the world, requiring 2,480,000 m³ of concrete, constructed from 1931 to 1936. (Courtesy of the United States Bureau of Reclamation.)

There are a number of approaches which may be taken. The simplest is to replace as much Portland cement as possible with other binder materials, such as fly ash, slag, and other pozzolanic materials. Fillers such as finely ground limestone may further reduce the Portland cement content. Using fly ash, up to 50% of the Portland cement may be replaced, though such cements cannot be used to produce very high strength concretes, or concretes that must have a high rate of early strength gain. Some blast furnace slags can be used at replacement rates of up to about 80%.

It has also been shown that, by very carefully controlling the particle size distribution of the concrete, and thereby optimizing the particle packing density (Fennis and Walraven, 2012), significantly lower cement contents can be used without adversely affecting the concrete properties. However, optimizing the particle packing for a given suite of materials can be time consuming (and costly), and very careful attention must be paid to quality control during production of the concrete. Because of the complexity of such mixes, they are very sensitive to just how they are produced and placed.

Much current research on concrete sustainability has focused on the use of various SCMs in conjunction with the appropriate chemical admixtures.

While this can certainly help to reduce CO_2 emissions, a more effective way of improving the sustainability of concrete is to make it more durable, that is, to increase the effective service life of the concrete structure. Ideally, this involves not only the design of the concrete mixture but also the design of the structure itself. Increasingly, specifications for concrete structures are beginning to contain clauses requiring a service life of 100 years, or even longer. Unfortunately, there is no general agreement on just what this means, on how to "guarantee" a specific service life, or on what tests should be carried out to confirm that the specified requirements have been met.

Building codes and standards are not particularly helpful in this regard. Service life is defined in rather vague terms, for example,

> CSA-S6-14 (2014): The actual period of time during which a structure performs its design function without *unforeseen* maintenance and repair.
> ACI 365 1R-00 (2000): Service life...is the period of time after...placement during which all the properties exceed the *minimum acceptable values* when routinely maintained.
> *fib* Bulletin No. 34 (2006): Design service life—assumed period for which a structure or a part of it is to be used for its intended purpose.

Contract documents are little more specific, for example,

> ...perform its design function without significant repairs, rehabilitation or replacement.
> Service life shall be defined as the time in service until spalling of the concrete occurs.

None of this really provides much guidance for the people charged with specifying the concrete or designing the structure. There is little or no attention given to the trade-offs between incremental initial costs to achieve durability and future costs of maintenance and rehabilitation. Indeed, while the level of maintenance and rehabilitation is crucial to predictions of service life, they are generally only vaguely defined. Finally, there is no real guidance on how to define or evaluate the "end of service life," since the tests and technologies that might be applied 100 years hence are largely unknown. One can foresee a lucrative future for lawyers and "experts" when these issues are litigated a century from now!

The first reinforced (with iron) concrete structure was a four-story house built in Paris in 1853 by François Coignet (Figure 1.6, Bajart, 2010), and this may provide a case in point for the difficulties alluded to above. What would have happened if Coignet had specified a 150-year service life for this house? How would it have been designed in the nineteenth century to conform to the requirements of the twenty-first century? It probably would

Figure 1.6 First house built with reinforced concrete, by Francois Coignet, Paris, 1853; now in a state of serious disrepair. (Courtesy of Bajart, E., 2010, Maison de Francois Coignet, available at: http://commons.wikimedia.org/wiki/File%3AMaison_Fran%C3%A7ois_Coignet_2.jpg, accessed on February 27, 2016.)

not be in the sorry state that it is in now if "significant repairs" had been carried out in a timely fashion, but how could they possibly have been able to specify this, and who would be responsible today for enforcing this specification?

The purpose of this book, then, is to try to elucidate the links between the behavior of concrete as a material, and the durability, sustainability, and cost of the structures for which it is to be used. Currently, a number of different organizations, such as ACI, RILEM, *fib*, and various national bodies are grappling with these issues, but there is as yet no common approach. It is hoped that this book may help to inform these discussions.

REFERENCES

ACI 365 1R-00, 2000, *Service-Life Prediction*, American Concrete Institute, Farmington Hills, MI.

Bajart, E., 2010, Maison de Francois Coignet, available at: http://commons.wikimedia.org/wiki/File%3AMaison_Fran%C3%A7ois_Coignet_2.jpg, accessed on February 27, 2016.

CSA-S6-14, 2014, *Canadian Highway Bridge Design Code*, Canadian Standards Association, Toronto, Canada.

Felice, M., 2013, Material of the month: Concrete, *Materials World*, Sept. 1, 3pp.

Fennis, S. A. A. M. and Walraven, J. C., 2012, Using particle packing technology for sustainable concrete mixture design, *Heron*, 57(2), 73–101.

fib Bulletin No. 34, 2006, *Model Code for Service Life Design*, Fédération Internationale du Beton (fib), Lausanne, Switzerland.

Hammond, G. P. and Jones, C. I., 2008, Embodied energy and carbon in construction materials, *Proceedings of the Institution of Civil Engineers—Energy*, 161(2), 87–98.

Scrivener, K. L., 2014, Options for the future of cement, *The Indian Concrete Journal*, 88(7), 11–21.

Thompson, S. E., 1909, *Concrete in Railroad Construction*, The Atlas Portland Cement Company, New York.

Chapter 2

Concrete as a modern construction material

2.1 OVERVIEW

The properties and structure of concrete have been thoroughly covered in various textbooks and reference books, addressing the engineering and scientific aspects (Mindess et al., 2003; Mehta and Monteiro, 2006; Aitcin, 2008; Aitcin and Mindess, 2011). The objective of this chapter is to provide a short overview, with particular emphasis on issues that are related to concrete durability, which will be referred to later in this book.

Concrete is essentially a composite material consisting of a particulate aggregate phase embedded in a continuous paste matrix, which binds the whole mass together. The overall properties of this composite depend on the characteristics of each of the components and the interaction between them at their interfaces. This is clearly demonstrated in the mechanical response of concrete (Hsu et al., 1963), in which each of the components is linear elastic while the overall concrete behavior is nonlinear (Figure 2.1). This nonlinearity is due to the formation of internal microcracks, which start at the aggregate–paste interface and propagate into the matrix as load is increased (Glucklich, 1968).

These microcracks may have a detrimental effect on durability as they can provide access to deleterious materials. However, their formation may also provide a positive influence by allowing for some relaxation in the stresses that develop upon concrete drying in restrained conditions. Shrinkage takes place both during the hardening of concrete and later on when it is exposed to dry environmental conditions. The cracks, if kept below about 100 μm in width, may not be detrimental, as the penetration through them is not much greater than that through the matrix (Figure 2.2).

When the crack is sufficiently small in width, it consists of surfaces that are tortuous and may come into contact with each other at some points, thus providing interlocking which can enable partial stress transfer across the crack. Therefore, the control of cracking through considerations of fracture mechanics and ductility enhancers, such as reinforcement by fibers or incorporation of polymers, can provide means for enhancing durability performance.

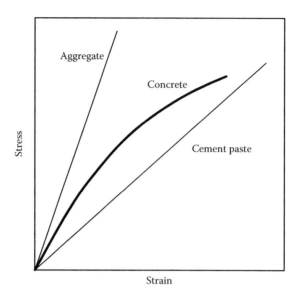

Figure 2.1 Schematic description of the stress–strain curve in compression of concrete and its components. (Based on concepts of Glucklich, J., 1968, *Proceedings of International Conference on the Structure of Concrete*, Cement and Concrete Association, London, UK, pp. 176–185.)

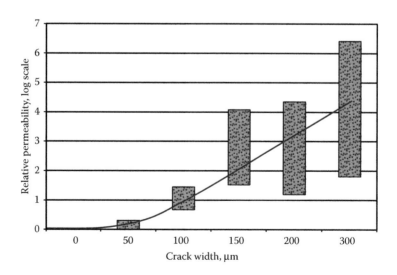

Figure 2.2 Effect of crack width on penetration relative to matrix penetration. (After Bentur, A. and Mitchell, D., 2008, Materials performance lessons, *Cement and Concrete Research*, 38, 259–272.)

If cracking is under control, the most important parameter for the mitigation of degradation processes is the density of the concrete matrix, which can be produced to be of low porosity. This is achieved by controlling its physical and chemical parameters, such as the water/cement (w/c) ratio, particle size distribution, and the chemical nature of the binder. The use of physical means alone to reduce w/c ratio to obtain a dense concrete can become a valuable tool for the control of deleterious chemical reactions such as sulfate attack (Figure 2.3). This approach can be used to eliminate the need for the use of cements of special chemical composition, such as sulfate-resistant cements.

The challenge of achieving a low w/c ratio lies in the ability to design concrete mixes with relatively low water content while preserving the workability of the fresh concrete. Thus, achieving durability control requires the use of rheological concepts to develop new workability control technologies and their successful incorporation in the mix design.

A modern materials science approach is currently used in its most comprehensive sense to control the fundamental properties of concrete and through it to provide tools for the design of durable concrete structures. Some of the more significant aspects will be reviewed here, addressing (i) hydration, paste matrix microstructure, and pore solution composition; (ii) control of the concrete composition and microstructure to provide

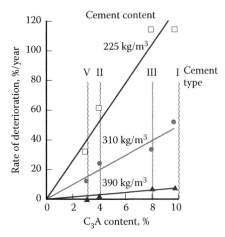

Figure 2.3 Effect of cement content and composition on the resistance to sulfate attack: higher cement content is associated with lower w/c ratio; for the 390 kg/m³ cement content, the resistance to sulfate attack of normal cement having high C_3A content is equivalent to that of sulfate-resistant cement having low C_3A content. (Adapted from Soroka, I., 1993, *Concrete in Hot Environment*, E&FN SPON; Based on Verbeck, G. J., 1968, Field and laboratory studies of the sulphate resistance of concrete, in *Performance of Concrete*, G. E. Swenson (ed.), University of Toronto Press, Toronto, Canada, pp. 113–124.)

mixes that are dense and workable; and (iii) developing systems in which deformations are under control and cracking is minimized.

2.2 HYDRATION, MICROSTRUCTURE, AND PORE SOLUTION COMPOSITION

2.2.1 Range of binder compositions

The active compounds in Portland cement, providing the strength and engineering characteristics to the concrete, consist mainly of calcium silicates produced in a kiln at high temperatures. The calcium silicate compounds can assume a range of compositions and their activity increases with the increase in the calcium content in the compound. In the case of Portland cement, the main two calcium silicate compounds are C_3S and C_2S (shorthand notation where C stands for CaO, S for SiO_2, and H for H_2O), and upon reaction with water (hydration reaction), they produce C–S–H gel (amorphous in nature) and CH, which is in a crystalline form, following the equation below:

$$(C_3S, C_2S) + H \rightarrow C-S-H + CH \tag{2.1}$$

The rate of reaction of the C_3S is higher than that of the C_2S, and the C–S–H formed has an average C/S ratio of about 1.5, which is similar for the hydration of both C_3S and C_2S. Thus, from stoichiometric considerations, the hydration of C_3S produces more crystalline CH than that of C_2S. The C–S–H gel in its amorphous form has a very high intrinsic surface area, which provides sites for strong physical interactions across the surfaces. This is the main mechanism by which the hydration products bind the whole mass of the matrix and the aggregates into the solid hardened concrete. The other principal phases in Portland cement, C_3A, and C_4AF (shorthand notation: C for CaO; A for Al_2O_3; and F for Fe_2O_3) also react with water. The C_3A reaction takes place early on at a very high rate, but the hydration products of both the C_3A and the C_4AF do not have significant binding properties.

The compounds of Portland cement form in the cement kiln at about 1450°C from a raw feed that consists largely of limestone, quartz, and clays. The range of compositions characteristic of Portland cement is shown in Figure 2.4 on the ternary $CaO-SiO_2-Al_2O_3$ diagram. They are positioned close to the CaO corner, where the compounds are richer in CaO and thus exhibit high hydraulic reactivity.

High-temperature compounds other than Portland cement, which are lower in CaO (shown in Figure 2.4), can be obtained in a variety of industrial processes, such as burning of coal for energy production and

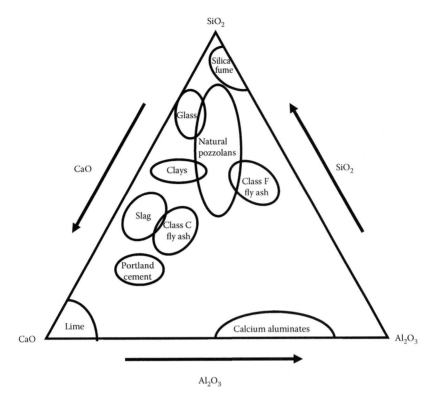

Figure 2.4 Ternary CaO–SiO₂–Al₂O₃ diagram showing the composition of Portland cement and other materials (mineral additives) that are incorporated with Portland cement in a variety of blends. (After Bentur, A., 2002, Cementitious materials—Nine millennia and a new century: Past, present and future, *ASCE Journal of Materials in Civil Engineering,* 14(1), 2–22.)

high-temperature treatment of iron ores to produce steel, in which the byproducts obtained are fly ash and blast-furnace slag (BFS), respectively. The range of their compositions is also described in the ternary diagram (Figure 2.4). Additional materials included in this category are silica fume and metakaolin, also produced by heat treatments, and natural pozzolans produced in volcanic eruptions. All of these have little or no reactivity with water on their own, but they can be activated by Portland cement or alkaline solutions. These nonhydraulic classes of materials are referred to generically as "mineral additives" or "supplementary cementitious materials." In concrete technology, these materials are usually mixed with Portland cement, either as components in blended cement (Table 2.1) or by direct addition to the concrete mixer. They react with the CH liberated by the Portland cement hydration to form a range of products. Most of these products are amorphous in nature and their composition is largely of calcium

Table 2.1 Compositions of Portland cement and blended cement specified in the EN 196 standard, based on replacing part of the Portland cement with mineral additives (supplementary cementitious materials)

Main types	Notation of products (types of common cements)		Composition (% by mass)ᵃ										
			Clinker	Blast-furnace slag	Silica fume	Pozzolana		Fly ash		Burnt shale	Limestone		Minor additional constituents
						Natural	Natural calcined	Siliceous	Calcareous				
			K	S	Dᵇ	P	Q	V	W	T	L	LL	
CEM I	Portland cement	CEM I	95–100	—	—	—	—	—	—	—	—	—	0–5
CEM II	Portland slag cement	CEM II A-S	80–94	6–20	—	—	—	—	—	—	—	—	0–5
		CEM II B-S	65–79	21–35	—	—	—	—	—	—	—	—	0–5
	Portland silica fume cement	CEM II A-D	90–94	—	6–20	—	—	—	—	—	—	—	0–5
	Portland pozzolana cement	CEM II A-P	80–94	—	—	6–20	—	—	—	—	—	—	0–5
		CEM II B-P	65–79	—	—	21–35	—	—	—	—	—	—	0–5
		CEM II A-Q	80–94	—	—	—	6–20	—	—	—	—	—	0–5
		CEM II B-Q	65–79	—	—	—	21–35	—	—	—	—	—	0–5
	Portland fly ash cement	CEM II A-V	80–94	—	—	—	—	6–20	—	—	—	—	0–5
		CEM II B-V	65–79	—	—	—	—	21–35	—	—	—	—	0–5
		CEM II A-W	80–94	—	—	—	—	—	6–20	—	—	—	0–5
		CEM II B-W	65–79	—	—	—	—	—	21–35	—	—	—	0–5
	Portland burnt shale cement	CEM II A-T	80–94	—	—	—	—	—	—	6–20	—	—	0–5
		CEM II B-T	65–79	—	—	—	—	—	—	21–35	—	—	0–5

(Continued)

Table 2.1 (Continued) Compositions of Portland cement and blended cement specified in the EN 196 standard, based on replacing part of the Portland cement with mineral additives (supplementary cementitious materials)

Main types	Notation of products (types of common cements)	Composition (% by mass)[a]										
		Clinker K	Blast-furnace slag S	Silica fume D[b]	Pozzolana Natural P	Pozzolana Natural calcined Q	Fly ash Siliceous V	Fly ash Calcareous W	Burnt shale T	Limestone L	Limestone LL	Minor additional constituents
	Portland limestone cement	CEM II A-L	80–94	–	–	–	–	–		–	6–20	–	0–5
		CEM II B-L	65–79	–	–	–	–	–		–	21–35	–	0–5
		CEM II A-LL	80–94	–	–	–	–	–		–	–	6–20	0–5
		CEM II B-LL	65–79	–	–	–	–	–		–	–	21–35	0–5
	Portland composite cement[c]	CEM II A-M	80–94					6–20					0–5
		CEM II B-M	65–79					21–35					0–5
CEM III	Blast-furnace cement	CEM III A	35–64	36–65	–	–	–	–	–	–	–	–	0–5
		CEM III B	20–34	66–80	–	–	–	–	–	–	–	–	0–5
		CEM III C	5–19	81–95	–	–	–	–	–	–	–	–	0–5
CEM IV	Pozzolanic cement[c]	CEM IV A	65–89	–		11–35				–	–	–	0–5
		CEM IV B	45–64	–		36–55				–	–	–	0–5
CEM V	Composite cement[c]	CEM V A	40–64	18–30	–	18–30				–	–	–	0–5
		CEM V B	20–39	31–50	–	31–50				–	–	–	0–5

a The values in the table refer to the sum of the main and minor additional components.
b The proportion of silica fume is limited to 10%.
c In Portland composite cements CEM II A-M and CEM II B-M, in pozzolanic cements CEM IV A and CEM IV B, and in composite cements CEM V A and CEM V B, the main constituents other than clinker shall be declared by designation of the cement.

silicate hydrate gel, C–S–H, having cementing properties similar to the C–S–H formed during Portland cement hydration, but usually of lower C/S ratio. These products can incorporate within them some Al and Fe ions.

2.2.2 Microstructure and pore solution composition

The specific volume of the hydration products is about twice that of the cement grains from which they have been formed. Thus, they have the ability to fill the space between the cement grains that was occupied by water, as shown schematically in Figure 2.5. In systems with higher w/c ratio, the spacing between the grains is larger, and as a result, not all of it is filled by hydration products and pores remain in the hardened materials, which is therefore more porous (Figure 2.5).

The more porous material is both weaker and more permeable. The basis of modern concrete technology is the control of strength and permeability by means of the w/c ratio, as shown in Figure 2.6.

The strength and permeability depend not only on the w/c ratio but also on the extent of hydration that needs to be as high as possible in order to materialize the potential of the concrete. This requires proper water curing until sufficiently dense structure is achieved, as demonstrated in Table 2.2.

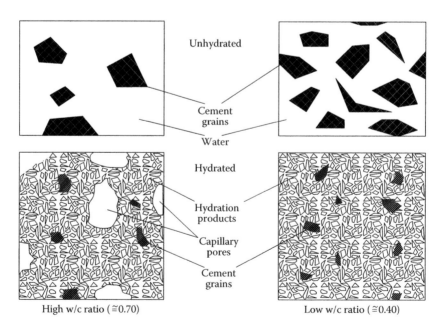

Figure 2.5 Schematic presentation of the pore filling of hydration products in different w/c ratio pastes. (After Bentur, A., Diamond, S., and Berke, N., 1997, *Steel Corrosion in Concrete*, E&FN SPON/Chapman and Hall, UK.)

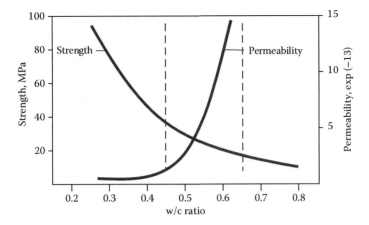

Figure 2.6 Effect of w/c ratio on strength and permeability.

The structure formed can be described in terms of pores and hydration products, largely C–S–H, which consists of nanometric size foils that are grouped together without any particular order, as shown schematically in Figure 2.7.

This simplified model can serve to calculate the volume of pores and their relative content for a given w/c ratio and degree of hydration; from there, an estimate of properties can be made in terms of the solid space ratio. The outcomes with regards to strength and permeability are shown schematically in Figure 2.8.

The spaces between the foil-like particles are small, in the range of 10 nm, and water is bound to their surfaces by physical forces. Drying of water from these spaces results in overall shrinkage, which is induced by forces of attraction between the surfaces of the foils. The attractive forces become larger as the shielding induced by the presence of H_2O molecules is reduced upon their removal by drying.

Table 2.2 Effect of duration of curing on the relative permeability of 0.70 w/c ratio cement paste with the progress of hydration

Age (days)	Relative coefficient of permeability
5	1.0000
6	0.2500
8	0.1000
13	0.0125
24	0.0025
Ultimate	0.0015

Source: Adapted from the data of Powers, T. C. et al., 1954, Permeability of Portland cement paste, *Journal of the American Concrete Institute, Proceedings*, 5, 285–298.

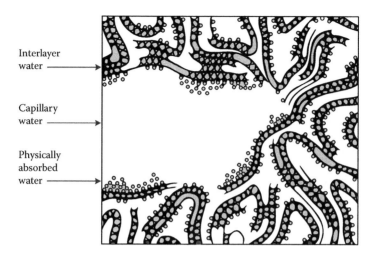

Interlayer water

Capillary water

Physically absorbed water

Figure 2.7 Schematic description of the structure of hydrated cement showing the C–S–H particles, interlayer water, and capillary pores. (After Bentur, A., 2002, Cementitious materials—Nine millennia and a new century: Past, present and future, *ASCE Journal of Materials in Civil Engineering*, 14(1), 2–22; Adapted from Feldman, R. F. and Sereda, P. J., 1968, A model for hydrated Portland cement paste as deduced from sorption length change and mechanical properties, *Materials and Structures*, 1, 509–520.)

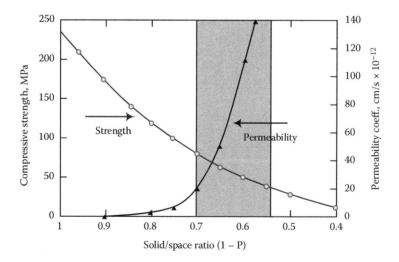

Figure 2.8 Solid/space ratio and relations with strength and permeability; the shaded area shows the typical values in hydrated cement paste. (After Mehta, P. K. and Monteiro, P. J. M., 2006, *Concrete Microstructure, Properties and Materials*, 3rd edition, McGraw-Hill, by permission of McGraw-Hill Education.)

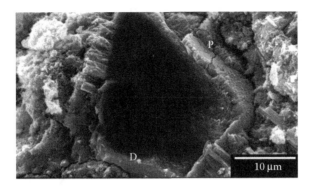

Figure 2.9 Hydrated calcium silicate grain showing the denser C–S–H formed within the original grain boundary (D) and the more porous C–S–H outside the grain boundary (P). (After Aitcin, P. C., 2008, *Binders for Durable and Sustainable Concrete*, Taylor & Francis.)

Modeling of the hydration products as a uniform and homogeneous medium is adequate for engineering-type calculations such as the ones used to obtain Figure 2.8. Yet, the assumption of homogeneity does not reflect the actual microstructure as observed in the electron microscope (Figure 2.9). Unhydrated grains can be seen as well as various types of C–S–H gel, with the denser C–S–H formed within the original boundaries of the reacting cement grain and the more porous C–S–H beyond its boundaries.

The nonuniform nature of the hydration products is especially evident at the early stages of hydration, during the first few days, when there is a considerable reaction of the aluminate compounds immediately after the start of the C_3S reaction (Figure 2.10).

The first few hours after the addition of water is the period during which the most rapid reactions occur. Since these are all exothermic reactions, their intensity is usually quantified in terms of the rate of heat liberation (Figure 2.11).

During the first few minutes after the mixing with water, there is dissolution of ions from the surface of the cement grains, without any appreciable reaction through the dormant period. At this stage, the pore water has become an electrolytic solution with mostly Ca^+, Na^+, and K^+ cations and OH^- anions; the solution is also rich in SO_4^- ions, and as long as they are present in solution, the C_3A hydration is delayed. When the solution becomes supersaturated with regard to Ca^+ ions, the C_3S starts to hydrate, as seen by the rise in heat liberation during the setting and early hardening reaction stage. This is the time at which setting takes place: sufficient hydration products are formed and start to fill the water-held spaces and to form initial contacts, which gradually turn the paste from a fluid to a rigid solid. The start of the heat liberation rise at this stage is usually

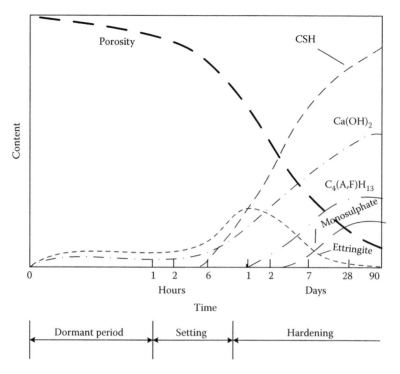

Figure 2.10 Schematic presentation of the hydration reactions from early age to maturity. (After Soroka, I., 1979, *Portland Cement Paste and Concrete*, The Macmillan Press, following personal communications with Locher and Richartz.)

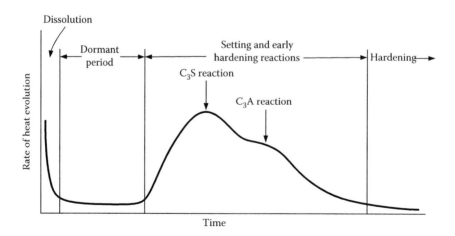

Figure 2.11 Schematic description of heat liberation curve of hydrating cement showing the different stages of reaction. (Adapted from Mindess, S., Young, J. F., and Darwin, D., 2003, *Concrete*, 2nd edition, Prentice-Hall.)

the time when initial setting takes place. The time at which the heat libera-tion peak is approached marks the final setting and the paste starts to gain strength, that is, early hardening. Beyond this time starts the hardening stage where the continuous hydration reaction leads to the buildup of the strength during the first few weeks of curing and beyond.

The C_3A reaction is delayed due to the presence of the SO_4^- ions, which react with C_3A to form a layer of ettringite on its surface, to retard its further reaction. When all of the SO_4^- has been depleted, the retarding influence of the ettringite layer is terminated and it gradually transforms to monosulfate while the C_3A enters into a rapid reaction. This occurs, however, after the C_3S has reacted significantly and thus cannot be interfered with by the C_3A reaction. This sequence of reactions can be clearly seen in Figure 2.10 with the reduction of the amount of ettringite and the formation of monosulfate.

Ettringite can reform at a later stage if a source of external SO_4^- ions becomes available. At this stage, when the matrix is hardened and rigid, the formation of the ettringite can be described in simplified terms as devel-oping with a morphology of long and slender crystals, pushing against a rigid material, to lead to the development of internal expansion stresses and to cracking. This is a simplistic description of the sulfate attack process, which can be eliminated by the use of low C_3A cements or concretes of low w/c ratio (Figure 2.3). (See also Chapter 4.)

The composition of the pore water in the concrete is quite important with regard to the control of durability. As stated above, the water is not "pure" but is rather an electrolytic solution in which the Ca^+, Na^+, and K^+ cations are balanced by OH^- anions, with a pH of about 13.5. This is rather a high value, indicative of a very high OH^- ion concentration. The pH is much higher than would be expected from a pore solution in contact with CH, where it is about 12.5. The elevated level is the result of the presence of the alkalis, coming largely from the cement. The high pH is beneficial from the point of view of the steel that is passivated when in contact with such a pore solution. It is also favorable for accelerating the activation of mineral additives by dissolution of their glassy phase. However, for the same reasons, it can trigger alkali–aggre-gate reactions when aggregates contain a reactive amorphous phase, which then reacts and forms a swelling type of gel. Loss of pH by carbonation can be detrimental for the protection of the steel, leading to its depassivation.

2.3 CONTROL OF THE RHEOLOGY OF FRESH CONCRETE

The production of concrete structures and components requires mixes that can be mixed, placed, and finished easily (i.e., minimum amount of work), while maintaining a homogeneous mix (i.e., not segregating). Traditionally, ease of flow was achieved by the presence of a sufficient content of water. However, the use of water for workability control poses a practical limit on the reduction of the w/c ratio to achieve concretes of low permeability and high durability (Figure 2.6). Also, enhancing the workability of fresh concrete (high consistency and improved flow) by adding water results usually in a segregation process, in which the coarse aggregates tend to settle downwards and a slurry of fines (rich in cement particles) tends to bleed upwards, resulting in inhomogeneity and compromising durability by creating voids and defects within the material. The properties of fresh concrete and their control are therefore a major factor in developing mixes of improved durability. The behavior of fresh concrete is governed by an amalgam of several characteristics such as ease of flow and resistance to segregation, collectively called "workability." Workability cannot be quantified in terms of a single basic parameter of physical significance. It can however be described by a flow curve, in terms of shear stress versus rate of shear strain, as shown schematically in Figure 2.12a. The curves for concrete can be modeled as a linear relation (Bingham body) and quantified in terms of two parameters: yield stress and plastic viscosity (Figure 2.12b). Extensive treatment of these issues and their characterization by rheological parameters can be found in several publications and books (e.g., Wallevik, 1998; Khayat and Wallevik, 2009; Wallevik and Wallevik, 2011).

Rheological curves are usually obtained with rotational viscometers, using special simple impellers for homogeneous materials (e.g., paste) or the ones with special geometry for nonhomogeneous fluids such as concrete that contains large aggregates.

Reduction of the water content in concrete without compromising workability (to obtain reduced w/c ratio without increasing the cement content) can be achieved by the use of water-reducing chemical admixtures. These admixtures cause the cement grains to repel each other, and this shows up in the rheological flow curves by reduction in the yield stress (improving the flow properties) without reducing the plastic viscosity (which counteracts the tendency for segregation and separation). The combined effect of adding the chemical admixture and reducing the water content (i.e., low w/c ratio concretes of good workability) leads to a reduction in the yield stress and an increase in the plastic viscosity (Figure 2.12b).

Extensive discussions of the structure and mode of operations of admixtures can be found in several references (e.g., Wallevik and Wallevik, 2011; Aïtcin and Flatt, 2016). The more traditional water-dispersing chemical admixtures were based on the adsorption of molecules on the cement grain surfaces, resulting in an electrical charge buildup, leading to repulsion between cement particles and enabling the breakup of the clumps of cement grains (Figure 2.13a). This is the mode of action of the conventional water-reducing admixtures and the first generations of the high-range water-reducing (HRWR) admixtures, which could lead to water reduction by about 20%. The more modern HRWR chemical admixtures, with higher water-reducing capacity, are based on the carboxylic polymer family, which enables the tailoring of co-polymer molecules having one branch adsorbed

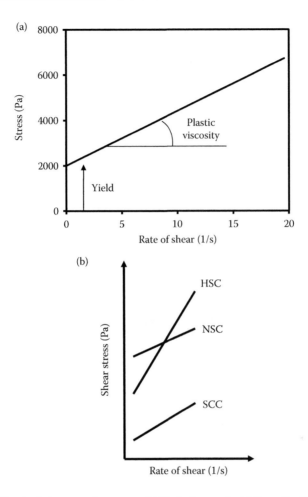

Figure 2.12 Rheological curves of concrete: (a) Typical curve that follows the Bingham body model can be described in terms of two parameters: yield stress and plastic viscosity. (b) Schematic presentation of rheological curves of NSC, HSC, and self-consolidating concrete (SCC). (From Bentur, A., 2002. *ASCE Journal of Materials in Civil Engineering*, 14(1), 2–22. With permission; (a) Adapted from Struble, L., Ji, X., and Dalens, G., 1998, Control of concrete rheology, in *Materials Science of Concrete: The Sidney Diamond Symposium*, M. Cohen, S. Mindess, and J. Skalny (eds.), The American Ceramic Society, pp. 231–245; (b) Adapted from Wallevik, O., 1998, *Rheological Measurements on Fresh Cement Paste, Mortar and Concrete by Use of Coaxial Cylinder Viscometer*, Icelandic Building Research Institute, Iceland.)

on the binder grain surface and a graft chain protruding outwards of the grain (Figure 2.13b). This structure and mode of interaction with the grain surface provides steric hindrance, which can be quite effective in keeping the cement grains apart and can induce very effective water reduction of 30% and more. These modes of action cause the cement grains to repel

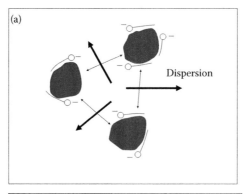

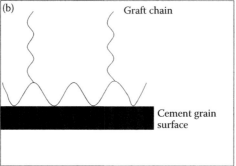

Figure 2.13 Schematic description of the action of water-reducing admixture (a) based on electrical charge buildup and (b) based on steric hindrance of the carboxylic-type HRWR showing its interaction with the surface of cement grain. (After Bentur, A., 2002. *ASCE Journal of Materials in Civil Engineering*, 14(1), 2–22; (a) Adapted from Young, J. F. et al., 1998, *The Science and Technology Civil Engineering Materials*, Prentice-Hall; (b) Adapted from Haneharo, S. and Yamada, K., 1999, *Cement and Concrete Research*, 29, 1159–1166.)

each other and counteract their tendency to clump together. The clumping is driven by the attraction induced by the positive and negative electrical charges on the surfaces of cement grains. This attraction is a hindrance to their ability to flow easily. Overcoming clumping by adding more water results in segregation. The use of the chemical admixtures eliminates the attraction forces without increasing the tendency for segregation.

2.4 CONTROL OF DENSITY AND CONCRETE MICROSTRUCTURE

The properties of the paste can serve as a reasonably good model to predict the behavior of concrete. However, there are differences which have considerable practical implications and some will be considered here:

- The microstructure of the paste in the vicinity of aggregates can be quite different from that in the bulk paste since there is a wall effect that does not allow the cement particles to pack efficiently at the interface. An additional influence is the bleeding effect, which results in some accumulation of water near the aggregate surfaces, particularly at the underside. All of these lead to a more porous transition zone around the aggregate, which can be a few tens of micrometers wide. This zone is a weak link in the concrete and provides sites for debonding, which starts at the interface upon loading and leads to nonlinear behavior of the stress–strain curve, as shown in Figure 2.1. The implications of the microcrack formation were discussed in the introduction to this chapter.
- In order to achieve very high strength and low permeability, a low w/c ratio is required, in the range of 0.2–0.3 (Figure 2.6). This can easily be obtained in paste but not in concrete, where there is a need for a water content in the range of 150–200 L/m^3 to obtain reasonable workability. For a 0.3 w/c ratio that would imply having almost 600 kg/m^3 of cement, which is not practical: (i) it will lead to cracking due to excessive shrinkage and thermal dilations, which are induced by the high cement content of the mix and (ii) it will result in a concrete that is very viscous and difficult to work with due to the very high content of fines.

The two issues identified above are just examples. Dealing with them requires attention at the level of the properties of the fresh concrete. This is where much of modern concrete technology has been focusing, to yield improvements in the hardened concrete which are related to the technologies which can be used to control its fresh behavior. The most efficient and important technologies are associated with (i) chemical admixtures which enable reduction in the content of mixing water (HRWR), (ii) use of mineral additives which can provide means for densifying the microstructure through pozzolanic reactions, and (iii) dense packing by having an optimal gradation of the particles of the binder phase (cement as well as mineral admixtures). The density of the microstructure obtained by these means can be estimated by the strength of the concrete. The evolution of these technologies over time and the strength level which they could generate are shown in Figure 2.14.

The combined incorporation of these new superplasticizers and mineral additives provides several interactions that are synergistic in nature, especially when the mineral additives consist of spherical particles, such as fly ash and microsilica:

1. Enhanced flow properties and additional reduction in water requirement by dispersion of the spherical particles in between the cement grains

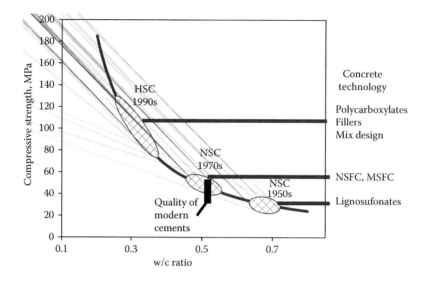

Figure 2.14 Evolution of technologies over time facilitating the production of concretes of enhanced density as shown by their w/c ratio and strength levels. (Adapted from Bentur, A., 2002. *ASCE Journal of Materials in Civil Engineering,* 14(1), 2–22.)

2. Additional densifying of the microstructure induced by the formation of more C–S–H gel by pozzolanic reactions (Figure 2.15)
3. Dense packing of the binder phase in the case of cement–microsilica–superplasticizer systems due to the gradation of the particles and their effective dispersion (Figure 2.16)
4. Dense packing of the binder phase at the aggregate surfaces in the case of cement–microsilica–superplasticizers systems, providing a strong bond with the aggregate to eliminate the weak link at the interface, which is characteristic of conventional concrete (Figure 2.17)

This microstructure results in a concrete with "true" composite action, in which the well-bonded aggregates can become reinforcing inclusions. The outcome is a concrete in which the strength can be higher than that of the matrix and the permeability is extremely low, due to the combined effect of dense packing of the matrix and the elimination of a porous aggregate–paste matrix interface. The result is a very strong concrete (Figure 2.18) of extremely low permeability (Aitcin, 1998).

The stress–strain curves in Figure 2.18 clearly indicate that the high-strength concrete (HSC) is more brittle (less post-peak carrying capacity relative to normal-strength concrete (NSC), which is due to the high bond strength. In the design of reinforced concrete structures, this drawback is usually taken care of by modification in the design of the reinforcing steel, in

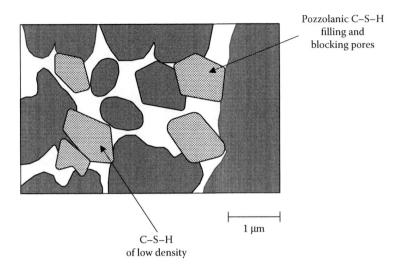

Pozzolanic C–S–H
filling and
blocking pores

1 μm

C–S–H
of low density

Figure 2.15 Additional densifying of the microstructure due to the pozzolanic reaction. (After Bentur, A., Diamond, S., and Berke, N., 1997, *Steel Corrosion in Concrete*, E&FN SPON/Chapman and Hall, UK.)

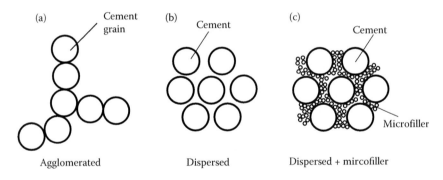

Figure 2.16 Schematic representation of cement particles: (a) in water flocculated state, (b) in dispersed state, and (c) optimal packing in dispersed state in combination with microsilica microfillers. (After Bentur, A., 2002, Cementitious materials—Nine millennia and a new century: Past, present and future, *ASCE Journal of Materials in Civil Engineering*, 14(1), 2–22, following concepts of Bache.)

particular the stirrups, which provide ductility characteristics to reinforced concrete structures.

A denser family of concretes has been developed in recent years, called ultrahigh-performance concrete (UHPC), which is based on the principles of the densely packed cement–microsilica binder system, in which the coarse aggregates are eliminated, enabling reductions of the water/binder ratio to values as low as 0.15. These systems are also characterized

(a)

(b)

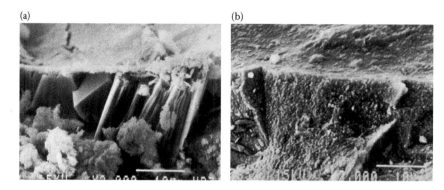

Figure 2.17 Interfacial microstructure in cement–microsilica–superplasticizer systems: (a) packing at the aggregate interface of cement paste system and (b) the microstructure in mature system with silica fume and superplasticizer. (After Bentur, A. and Cohen, M., 1987, The effect of condensed silica fume on the microstructure of the interfacial zone in Portland cement mortar, *Journal of the American Ceramic Society*, 70(10), 738–742.)

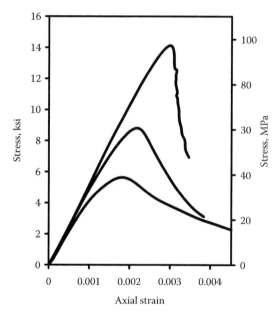

Figure 2.18 Stress–strain curves in compression of normal-, medium-, and high-strength concretes. (After Jansen, D. C., Shah, S. P., and Rossow, E. C., 1995, Stress-strain results of concrete from circumferential strain feedback control testing, *ACI Materials Journal*, 92(4), 419–428.)

by superior flow properties and allow for the incorporation of several percentages of steel fibers by volume, resulting in a composite with extremely dense microstructure, high compressive strength, and considerable ductility and flexural strength (Figure 2.19). These composites are drastically different in nature from other concretes and they are preferably applied as thin-sheet components. Their use has been demonstrated in lightweight bridge construction. They can also be incorporated with conventional concrete components by applying them as external layers to strengthen the concrete members and provide environmental protection. In these

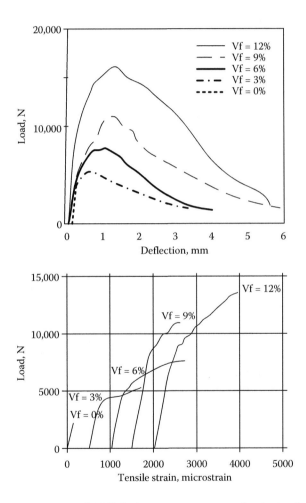

Figure 2.19 Flexural curve of UHPC demonstrating strain hardening behavior (i.e., increase in load-bearing capacity after the first cracking at the limit of proportionality). (After Tjiptobroto, P. and Hansen, W., 1993, Tensile strain hardening and multiple cracking in high-performance cement-based composites containing discontinuous fibers, *ACI Materials Journal*, 90, 16–25.)

applications, they have merit as repair materials and coatings to enhance durability.

2.5 DEFORMATIONS AND CRACK CONTROL

The internal structure of the hydration products and the exothermic nature of the hydration reactions result in deformations in the concrete, which are associated with interactions with the surrounding environment:

1. Time-dependent deformations resulting from slow migration of water from the pores and the interlayer spaces between the C–S–H particles, associated with wetting and drying induced by external environmental changes.
2. Thermal movements due to the differences in temperatures, induced by the heat generated during the hydration reactions and its subsequent dissipation to the surrounding environment.

These deformations may generate internal cracking if they occur in a concrete member that is under restrained conditions.

Loss of water due to drying is associated with gradual emptying of pores of different size ranges:

1. Bigger pores in which the water is very loosely held
2. Smaller pores where capillary forces are generated and the water is not free to move
3. Spaces between the C–S–H particles (sometimes referred to as interlayer water), ~10 nm wide, where water is tightly held to the surface of the particles

   In the capillary range, a meniscus is formed resulting in capillary compressive stresses in the solid structure, leading to shrinkage as water is removed. The removal of water from the interlayer spaces results in even greater shrinkage strains as the surfaces of the gel particles are being drawn together upon the removal of the water between their spaces.

Shrinkage strains can be generated due to internal drying in sealed concrete, when the water in the paste matrix is consumed in the hydration reactions. This phenomenon, known as autogenous shrinkage, is the result of the fact that the volume of the hydration products formed is smaller than the total volume of the reactants (cement grain + water). If this volume contraction (labeled as chemical contraction) is not balanced by external water, then internal drying (desiccation) will occur, leading to early-age shrinkage. This is particularly the case in low w/c ratio systems, where the total volume of water is small and thus the chemical contraction is relatively high.

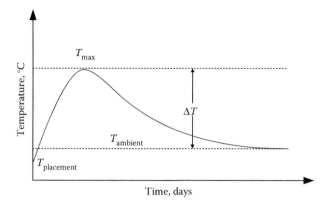

Figure 2.20 Temperature change during the first few days of hydration which leads to thermal shrinkage. (After Mehta, P. K. and Monteiro, P. J. M., 2006, *Concrete Microstructure, Properties and Materials*, 3rd edition, McGraw-Hill, by permission of McGraw-Hill Education.)

Additional deformation associated with the hydration process is caused by the exothermic nature of the reactions, which liberates heat intensively at early ages when the reaction rates are high (Figure 2.11). The result is an increase in the temperature during the first few days, followed by cooling down as the rate of heat dissipation to the surrounding environment exceeds the heat buildup due to the continued slow hydration (Figure 2.20).

The maximum temperature is reached within a few days; thereafter, the cooling takes place in a concrete mass that has become hard, leading to contraction. This contraction should be added to that which may occur due to shrinkage and the overall length change should be considered. The overall contraction which typically occurs in concretes at early ages can result in the buildup of tensile stresses in a restrained component. The extent of this buildup depends on the level of restraint, the modulus of elasticity at this early age, and the creep which releases some of the stress. If the tensile stress buildup during this time exceeds the development of the tensile strength, cracking may occur, as shown schematically in Figure 2.21.

Cracking of this kind can affect the durability performance by creating pathways for the penetration of deleterious materials into the concrete, and it often occurs when restraining of deformations occurring during the service life of a structure arise. Such deformations can be the result of plastic shrinkage and settlement, early thermal contraction, long-term drying shrinkage, and deformations induced during the corrosion of steel and alkali–silica reaction. The nature of the restraint and the type of cracking that is induced under such conditions are dealt with more detail in Chapter 4 and presented schematically in Figures 4.7 and 4.8.

Mitigation of shrinkage and cracking can be achieved through proper design of the concrete mix. Concrete is a composite material in which only

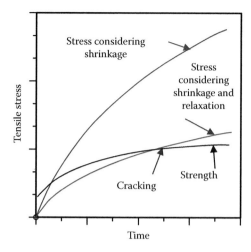

Figure 2.21 Schematic presentation of the processes involved in the buildup of tensile stresses in an early-age restrained concrete member due to its contraction during this time period.

the paste phase undergoes intense deformations; the overall behavior of the concrete is influenced by the aggregate content that exerts a mitigating effect. Thus, the design of the concrete to maximize the aggregate content by means such as optimal grading is required.

REFERENCES

Aitcin, P. C., 1998, *High-Performance Concrete*, E&FN SPON, London and New York.

Aitcin, P. C., 2008, *Binders for Durable and Sustainable Concrete*, Taylor & Francis, London and New York.

Aïtcin, P.-C. and Flatt, R. J. (eds.), 2016, *Science and Technology of Concrete Admixtures*, Woodhead Publishing, Cambridge, UK, 613pp.

Aitcin, P. C. and Mindess, S., 2011, *Sustainability of Concrete*, SPON Press, Oxon, UK.

Bentur, A., 2002, Cementitious materials—Nine millennia and a new century: Past, present and future, ASCE *Journal of Materials in Civil Engineering*, 14(1), 2–22.

Bentur, A. and Cohen, M., 1987, The effect of condensed silica fume on the microstructure of the interfacial zone in Portland cement mortar, *Journal of the American Ceramic Society*, 70(10), 738–742.

Bentur, A., Diamond, S., and Berke, N., 1997, *Steel Corrosion in Concrete*, E&FN SPON/Chapman and Hall, London, UK.

Bentur, A. and Mitchell, D., 2008, Materials performance lessons, *Cement and Concrete Research*, 38, 259–272.

Feldman, R. F. and Sereda, P. J., 1968, A model for hydrated Portland cement paste as deduced from sorption length change and mechanical properties, *Materials and Structures*, 1, 509–520.

Glucklich, J., 1968, *Proceedings of International Conference on the Structure of Concrete*, Cement and Concrete Association, London, UK, pp. 176–185.

Haneharo, S. and Yamada, K., 1999, Interaction between cement and admixture from point of cement hydration, adsorption behavior and paste rheology, *Cement and Concrete Research*, 29, 1159–1166.

Hsu, T. T. C., Slate, F. O., Sturman, G. M., and Winter, G. 1963, Microcracking of plain concrete and the shape of the stress-strain curve, *Journal of the American Concrete Institute, Proceedings*, 60(14), 209–224.

Jansen, D. C., Shah, S. P., and Rossow, E. C., 1995, Stress-strain results of concrete from circumferential strain feedback control testing, *ACI Materials Journal*, 92(4), 419–428.

Khayat, K. H. and Wallevik, O., 2009, *Workability and Rheology of Flowable and Self-Consolidating Concrete*, CRC Press, Boca Raton, FL, 256pp.

Mehta, P. K. and Monteiro, P. J. M., 2006, *Concrete Microstructure, Properties and Materials*, 3rd edition, McGraw-Hill, New York.

Mindess, S., Young, J. F., and Darwin, D., 2003, *Concrete*, 2nd edition, Prentice-Hall, Upper Saddle River, NJ, USA.

Powers, T. C., Copeland, L. E., Hayes, J. C., and Mann, H. M., 1954, Permeability of Portland cement paste, *Journal of the American Concrete Institute, Proceedings*, 51, 285–298.

Soroka, I., 1979, *Portland Cement Paste and Concrete*, The Macmillan Press, London, England.

Soroka, I., 1993, *Concrete in Hot Environment*, E&FN SPON, London.

Struble, L., Ji, X., and Dalens, G., 1998, Control of concrete rheology, in *Materials Science of Concrete: The Sidney Diamond Symposium*, M. Cohen, S. Mindess, and J. Skalny (eds.), The American Ceramic Society, Westerville, OH, pp. 231–245.

Tjiptobroto, P. and Hansen, W., 1993, Tensile strain hardening and multiple cracking in high-performance cement-based composites containing discontinuous fibers, *ACI Materials Journal*, 90, 16–25.

Verbeck, G. J., 1968, Field and laboratory studies of the sulphate resistance of concrete, in *Performance of Concrete*, G. E. Swenson (ed.), University of Toronto Press, Toronto, Canada, pp. 113–124.

Wallevik, O., 1998, *Rheological Measurements on Fresh Cement Paste, Mortar and Concrete by Use of Coaxial Cylinder Viscometer*, Icelandic Building Research Institute, Iceland.

Wallevik, O. H. and Wallevik, J. E., 2011, Rheology as a tool in concrete science: The use of rheographs and workability boxes, *Cement and Concrete Research*, 41, 1279–1288.

Young, J. F., Mindess, S., Gray, R. J., and Bentur, A., 1998, *The Science and Technology Civil Engineering Materials*, Prentice-Hall, NJ.

Chapter 3

Materials for concretes in relation to durability

3.1 INTRODUCTION

Prior to the early 1930s, concrete was a relatively simple, straightforward material, consisting only of "pure" Portland cement, aggregate, and water. These were combined in varying proportions to produce concretes with a range of compressive strengths and workabilities. In North America, it was only then that the first concrete additives began to come into general use: air-entraining, set-controlling, and water-reducing admixtures, and fly ash as a cement replacement. In Europe, ground-granulated blast-furnace slag (GGBFS) had been used since the beginning of the twentieth century, but the chemical admixtures were still not available until the 1930s. More recently, the numbers and types of materials being added to Portland cement and concrete have been growing at an accelerating pace, driven by a combination of economic considerations and a desire to make the cement and concrete industries more sustainable. Of course, both durable and non-durable concretes are made using much the same materials. The object is to ensure that the right materials are chosen for any particular project, taking into consideration both the exposure conditions and any special requirements for the concrete. In what follows, the materials found in modern concretes will be examined primarily in terms of their impact on the durability of concrete. This chapter deals only with the concrete-making materials themselves; the corrosion of steel reinforcement is dealt with in Chapters 4 and 8, and other durability issues are described in greater detail elsewhere in this book.

3.2 PORTLAND CEMENTS AND OTHER BINDERS

Of all of the materials used to make concrete, it is the Portland cement-based binder that is the most unstable part of the system, and is responsible for most of the durability problems that may beset concrete. It may be subject to both chemical and physical attack, as shown in Table 3.1. (A much more detailed list of deterioration mechanisms is provided in Chapter 4,

Table 3.1 Durability of Portland cement-based binders

Chemical attack
Leaching and efflorescence
Sulfate attack
Acids and alkalis
Alkali–aggregate reactions
Corrosion of steel and other metals
Seawater
Physical attack
Freezing and thawing
Wetting and drying
Thermal effects
Abrasion and wear

Table 4.1.) These durability issues are inherent with the use of Portland cements; they are functions of the physical and chemical nature of these cements. Fortunately, we already know how to deal with most of these problems.

It should be noted that in North America, where the bulk of cement production falls within the parameters of the five basic ASTM cement types, long-term performance data are readily available for both strength and durability. However, in Europe, the classification of EN cements (BS EN 197-1, 2011) presents a bewildering array of possibilities, as already shown in Table 2.1. The vast majority of cements on the European market are blended, often heavily so. Except for strength, there is little performance data on these "newer" cements, and so their long-term durability remains somewhat uncertain.

Of course, the first line of defense is to keep the water/binder (w/b) ratio as low as possible (and to ensure that the concrete is properly cured). This will greatly reduce the permeability of the concrete, thus inhibiting the ingress of deleterious chemicals into the interior of the concrete. Further reductions in permeability can be achieved by adding supplementary cementing materials (SCMs). In particular, silica fume leads to a reduction in both total porosity and the size of the pores. While these measures cannot eliminate chemical attack, they can slow it down considerably, thereby increasing the effective life of the concrete.

3.2.1 Chemical attack

In most cases, not much can be done with the cementing materials themselves to eliminate leaching and efflorescence, or chemical attack from acids or from water containing significant amounts of industrial or agricultural wastes. These types of problems are governed by the chemistry of the cements and of the deleterious chemicals. As stated above, these reactions can be delayed only by using a low w/b and the use of certain SCMs. However, there are some instances in which altering the cement (or binder) chemistry can be highly beneficial.

3.2.1.1 Alkali–aggregate reaction

For instance, where alkali–aggregate reaction (AAR) is likely to occur, reducing the alkali equivalent of the cement to below 0.6%, limiting the total alkali content of the concrete, and using a low w/b ratio will generally mitigate the problem (Blight and Alexander, 2011), or at least considerably diminish the rate of deterioration.

3.2.1.2 Sulfate attack

The mechanisms involved in sulfate attack are complex, and have been described in detail by Skalny et al. (2002). The requirements to mitigate this attack, however, are quite straightforward. Basically, where sulfate attack is expected, the use of a low C_3A cement is required. This must be used in combination with a low w/b ratio to be effective. Typical requirements for sulfate-resistant concrete based on the Canadian standard (CSA A23.1-14, 2014) are given in Chapter 9 (Table 9.1).

In North America, there are basically two common sulfate-resistant cements:

ASTM Type II: Moderate sulfate resistance, with $C_3A \leq 8\%$
ASTM Type V: High sulfate resistance, with $C_3A \leq 5\%$

These C_3A requirements can most often be met by using a blended cement, where the C_3A content is "diluted" by substituting an SCM for a portion of the Portland cement. Equivalence in sulfate resistance performance can be established between concretes made with sulfate-resistant cements and concretes made with normal Portland cement but with a lower w/b ratio, as shown for example in Chapter 2, Figure 2.3.

The situation is rather more complex in Europe, where blended cements are much more common than they are in North America. The current EU standard, EN-197-1 (2011) defines seven sulfate-resistant common cements (SR cements):

Sulfate-resisting Portland cement	
CEM I-SR0	C_3A content of clinker $= 0\%$
CEM I-SR3	C_3A content of clinker $\leq 3\%$
CEM I-SR5	C3A content of clinker $\leq 5\%$

Continued

Sulfate-resisting blast-furnace cement	
CEM III/B-SR	No requirement on C_3A
CEM III/C-SR	content of clinker

Sulfate-resisting pozzolanic cement	
CEM IV/A-SR	C_3A content of clinker $\leq 9\%$
CEM IV/B-SR	C_3A content of clinker $\leq 9\%$

Table 3.2 Sulfate-resisting common cements

Main types	Types of sulfate-resisting common cement	Composition (% by mass)				
		Main constituents				
		Clinker (K)	Blast-furnace slag (S)	Pozzolans (natural) (P)	Siliceous fly ash (V)	Minor constituents
CEM I	CEM I-SR0					
	CEM I-SR3	95–100				0–5
	CEM I-SR5					
CEM III	CEM III/B-SR	20–34	66–80			0–5
	CEM III/C-SR	5–19	81–95			0–5
CEM IV	CEM IV/A-SR	65–79		← ← 21–35 → →		0–5
	CEM IV/B-SR	45–64		← ← 36–55 → →		0–5

Source: Adapted from BS EN 197-1:2011, 2011, *Cement Composition, Specifications and Conformity Criteria for Common Cements*, British Standards Institution, London, UK.

The compositions of these SR cements are given in Table 3.2.

3.2.1.3 Acid attack

Concrete will be attacked by acids, the severity of the attack depending on the type of acid and its concentration. The $Ca(OH)_2$ in the hydrated cement is particularly susceptible to acid attack, but the C–S–H may also be attacked and, in some circumstances, the aggregate as well. The attack will occur when the pH of the liquid is below 6.5, becoming increasingly severe as the pH drops. The attack may be mitigated (but not eliminated) by blending the Portland cement with SCMs at replacement levels up to about 60% (Tamimi, 1997), such as fly ash, GGBFS, silica fume, or metakaolin, since this effectively reduces the amount of $Ca(OH)_2$ (Torii and Kawamura, 1994; de Belie et al., 1996; Kim et al., 2007).

3.2.1.4 Other considerations

It should be noted that modern concretes, and particularly high-performance concretes, are now much less "forgiving" than they used to be. That is, they are much more sensitive to exactly which raw materials are used, how they are mixed together, and how the concrete is placed and cured. In part, this is related to changes in the chemistry and fineness of grinding of the Portland cement, which in turn are being driven by the desire of contractors and owners to accelerate the rate at which concrete gains strength (Aïtcin and Mindess, 2015). This is particularly true for binders containing large quantities of fly ash or similar pozzolanic materials, which would otherwise have a somewhat slower rate of strength gain because the pozzolanic reaction is relatively slow. There are two principal ways of "speeding up" the hydration reactions: finer grinding of the Portland cement and increasing its C_3S and C_3A contents.

3.2.1.5 Fineness

The fineness of Portland cement has steadily increased over the past 50 years (Bentz et al., 2011). They found that the average fineness increased from about 330 m²/kg in the early 1950s to about 380–400 m²/kg in 2010. This has a number of consequences. Of course, the hydration rate increases, leading to a higher rate of strength gain at early ages, but at the cost of a lower ultimate strength. As well, the higher reaction rate leads to a faster rate of heat release, which can lead to a higher temperature buildup in large sections, and the possibility of thermal cracking. The higher cement fineness can also lead to more shrinkage cracking, and a reduced capacity for self-healing. Combined, these effects will lead to potentially less durable concrete.

3.2.1.6 C_3A content

From the point of view of the *cement* producer, producing a Portland cement with C_3A and C_4AF contents of about 8% each is close to ideal. At these contents, the clinker output of the kiln is maximized, and the kiln requires relatively little attention from the operator. Also, in the burning zone, the raw meal is easily transformed to clinker, because there is a good balance between the silicate phases and the interstitial (aluminate) phases. However, when producing a low C_3A clinker, the interstitial phase becomes too fluid in the burning zone of the kiln, requiring a reduction in the speed of rotation of the kiln, and hence a reduced output. The kiln also becomes more difficult to operate.

Conversely, from the point of view of the *concrete* producer, the C_3A is highly undesirable, because of its negative effects on the rheology and durability of the concrete. C_3A is the most reactive of the cement minerals and combines with the gypsum ($CaSO_4 \cdot 2H_2O$) to form ettringite ($6CaO \cdot Al_2O_3 \cdot 3SO_3 \cdot 32H_2O$), mostly in the form of needles. These ettringite needles decrease the workability; the higher the C_3A content, and the higher the fineness, the greater the loss in workability.

The C_3A may also lead to cement–superplasticizer incompatibility in low w/b ratio concretes. Finally, a high C_3A content may lead to durability problems, because ettringite is not stable in the high pH of the interstitial water in the capillary pores. After a few hours of hydration, some of the ettringite will be transformed into monosulfoaluminate ($3CaO \cdot Al_2O_3 \cdot CaSO_4 \cdot 12H_2O$). If the concrete is subsequently exposed to sulfates, both the ettringite and the monosulfoaluminate are unstable; this can cause durability problems such as delayed ettringite formation (DEF), sulfate attack, and reduced resistance to freeze–thaw cycles. This suggests that to produce durable high-performance concretes, cements with a C_3A content of about 6%, not too finely ground (300–350 m^2/kg), should ideally be used.

3.2.2 Physical attack

As with chemical attack, the first line of defense against physical attack is also to use as low a w/b ratio as is practicable for any individual structure. There are no particular "materials" solutions to problems arising from wetting and drying cycles. For abrasion and wear issues (see Chapter 9, Section 9.2.5), these are governed mostly by the aggregate, which is almost always harder and stronger than the cement paste. Stronger aggregates and a better paste–aggregate bond can only delay wear and abrasion, but cannot completely eliminate it (Mindess and Aïtcin, 2014).

3.2.2.1 Thermal effects

Mature hardened cement paste contains an appreciable amount of water: evaporable water in the larger capillary pores, the volume of this water depending upon the original water/cement (w/c) ratio and the degree of hydration; water physically adsorbed on the surface of the hydrated cement paste; and chemically combined water (i.e., the water involved in the hydration reactions). As the hydrated cement is heated from room temperature, the evaporable water begins to be driven off, with most of it lost by the time that the temperature reaches about 105°C. This is accompanied by considerable shrinkage of the paste. On further heating, by about 215°C, the remaining capillary water and some of the chemically bound water are lost (Phan et al., 2001). Between 300°C and 400°C, microcracking due to differential thermal deformations of the various phases in hydrated Portland cement occurs around the $Ca(OH)_2$ (portlandite) crystals, and around the larger still unhydrated cement grains (Piasta, 1984). Dehydration of the portlandite crystals begins at about 400°C, and decomposition is complete at about 600°C. The hardened cement paste continues to lose water continuously as the temperature increases, and is completely dehydrated by about 850°C. This is accompanied by a large increase in total porosity and

Table 3.3 Types of refractory concrete

Type	Service temperature limit (°C)	Cement	Aggregate
Structural	300	Portland	Common rocks
Heat resisting	1000	High alumina	Fine-grained basic igneous rocks (basalt, dolerite); heat-treated porous aggregates (bricks, scoria, expanded shale)
Ordinary refractory	1350	High alumina	Firebrick
Super-duty refractory	1450		Firebrick
	1350–1600	High alumina	Refractory (magnesite, silicon carbide, bauxite, fired clay)

Source: From Mindess, S., Young, J. F., and Darwin, D., 2003, *Concrete*, 2nd edition, Prentice-Hall, Upper Saddle River, NJ, 644pp. Copyright 2003. Reprinted by permission of Pearson Education, Inc., New York, NY.

permeability (Piasta et al., 1984; Lee et al., 2008; Liu et al., 2008). Some of the cement begins to melt above 1200°C, and by 1300–1400°C, it is essentially entirely melted.

This thermal degradation of the cement paste is essentially independent of the w/c ratio and the type of Portland cement. For extended exposure to very high temperatures (>1000°C), calcium aluminate or high-alumina cements should be used. Generally, the effects of high temperature are governed mostly by the aggregate properties; for long-term exposure to high temperatures, it would be best to use refractory aggregates (Table 3.3).

3.3 AGGREGATES

Generally, aggregates are harder and more durable than the matrix phase in concrete. However, there are some cases in which the aggregates either contribute directly to concrete deterioration or are susceptible to damage themselves under certain exposure conditions. Concrete deterioration mechanisms relating to aggregates have been reviewed in some detail by Alexander and Mindess (2005). They divide these mechanisms into two types: *intrinsic* mechanisms refer to processes that occur internally as a consequence of the nature of the constituent materials, and *extrinsic* mechanisms whereby concrete deteriorates from the actions of external agents. Some of the deterioration mechanisms relating to aggregates in concrete are given in Table 3.4 (taken from Alexander and Mindess, 2005), and are briefly discussed below.

Table 3.4 Deterioration mechanisms relating to aggregates in concrete

	Physical mechanisms	*Chemical mechanisms*
Intrinsic mechanisms	Dimensional incompatibility	Alkali–aggregate reaction
	• Thermal effects	Sulfides in aggregates
	• Moisture effects	Thaumasite sulfate attack
Extrinsic mechanisms	Freeze–thaw	Acid attack
	Surface wetting and drying (moisture cycles)	Alkali attack
	Surface abrasion and erosion	Other aggressive chemicals, e.g., sulfates

Source: From Alexander, M. and Mindess, S., 2005, *Aggregates in Concrete*, Taylor & Francis, Abingdon, UK, 435pp.

3.3.1 Physical mechanisms

Since concrete is a composite material consisting of aggregate particles embedded in a binder, we must consider the possibility of dimensional incompatibility between these two phases, due to either thermal or moisture effects.

3.3.1.1 Thermal effects

The coefficients of thermal expansion for aggregates and cements are quite different. For rocks (Alexander and Mindess, 2005), these range from about 1×10^{-6} to $14 \times 10^{-6}/°C$; for cement, they typically range from about 11×10^{-6} to $16 \times 10^{-6}/°C$. Siliceous minerals such as quartz ($12 \times 10^{-6}/°C$) have thermal coefficients about twice those of calcareous aggregates and certain granites (about $6 \times 10^{-6}/°C$). The coefficient of thermal expansion of concrete made with siliceous aggregates is thus significantly higher than that of concrete made with calcareous aggregates. Generally, this disparity in thermal expansion coefficients does not cause problems, but care should be taken when aggregates with low thermal coefficients are used in situations where fire resistance is required, or where there is cyclic freezing and thawing. In these cases, cracks may develop between the aggregate and the matrix, reducing the durability. It has been suggested (Callan, 1952) that the difference between the thermal coefficients of coarse aggregate and the mortar should not exceed $5 \times 10^{-6}/°C$.

The effect of temperature on aggregate properties becomes particularly important at high temperatures. Typical concrete aggregates are all thermally stable up to at least 350°C. At higher temperatures, their behaviors diverge. Aggregates that contain water will dehydrate, leading to a certain amount of shrinkage. The transformation of α-quartz to β-quartz at 573°C is accompanied by a volume increase of about 5.7%. At temperatures above about 600°C, some calcareous aggregates (calcite, dolomite, magnesite) will dissociate to form CO_2 and an oxide. In the temperature range 1200–1300°C, some igneous rocks show degassing and expansion.

The temperatures at which aggregates melt also vary widely, from basalt at about 1060°C to quartzite at about 1700°C. Thus, the thermal behavior of concrete is very sensitive to the precise nature of the aggregate(s) used, particularly since aggregates make up about 70% of the volume of high-performance concretes.

The coefficient of thermal expansion of cement paste is strongly dependent upon its moisture content. On the other hand, the thermal coefficients of most rocks depend on the temperature, with the coefficients increasing significantly as the temperature increases. Thus, at lower temperatures, the coefficient of thermal expansion of cement paste is higher than that of most rocks; at high temperatures, the reverse is true. What is clear is that as the temperature increases from ambient to over 1200°C, the differences in thermal coefficients of the cement and the aggregates, and the other volume changes that take place, will lead to considerable cracking (both micro and macro cracks) in the concrete. Of course, nothing can be done about the differences in thermal expansion coefficients between cement and aggregate.

3.3.1.2 Moisture effects

In concrete mix design, it is assumed that the aggregates are dimensionally stable. However, this is not always the case; there are aggregates that exhibit shrinking and swelling behavior on drying and wetting, and this can lead to very high values of concrete drying shrinkage, typically in excess of 1000×10^{-6}. This may lead to excessive deflections and cracking in structural elements. This phenomenon is well known for lightweight aggregates, but is also found with certain basic igneous rocks, some metamorphosed shales and slates, certain mudstones and sandstones, and other rock types as well. If these rocks are weathered, they should be considered as highly shrinkable, unless they are proven to be otherwise, either by tests or by a satisfactory service record. The best predictor of aggregates causing excess concrete shrinkage is aggregate absorption, with values above 0.8% indicating a much increased risk of excess drying shrinkage (Edwards, 1966).

3.3.1.3 Freezing and thawing

Certain types of aggregate may also suffer freeze–thaw damage; such aggregates typically have a total pore volume (or total absorption) greater than 1.5%. The rocks most susceptible to this type of damage are those with fine pores, a high total porosity, and a low permeability; this can occur in some cherts and shales.

3.3.1.4 Abrasion and erosion

The abrasion resistance of aggregates themselves is of little practical importance for most concrete construction. The exception is for those concretes exposed to surface wear, as in pavements, airport runways, dam spillways, concrete canals carrying silt or gravel, or abrasion-resistant floor toppings. Abrasion of the aggregate in concrete can occur only if the aggregate particles become exposed due to wearing away of the original concrete surface. In all cases, it is best to use aggregates that are hard, strong, and free of soft or friable particles; a low w/c ratio, high strength concrete is also useful. However, whatever the case, it is more useful to assess the concrete itself for its wear resistance since this takes into account both the aggregate properties and the paste–aggregate bond.

The type of abrasive wear does impose somewhat different aggregate requirements. For *erosion* due to suspended solids in flowing water, larger aggregate particles work better, while for concrete subjected to *cavitation*, smaller (<20 mm) particles are preferred. For *abrasion*, aggregate behavior becomes increasingly important as strengths fall further and further below about 40 MPa.

There are no very good tests for the abrasion resistance of aggregates. Probably, the most common test is the Los Angeles abrasion test (ASTM C131 and C535), which involves ball milling a sample of aggregate and measuring the mass loss passing a 1.70-mm sieve. This test, however, does not correlate well with concrete behavior in the field; it is a better measure of the tendency of aggregates to break down during handling.

3.3.1.5 Maximum aggregate size

Although the maximum size of the coarse aggregate is not in itself a durability problem, it should be noted that there has been a tendency to go to smaller maximum aggregate sizes, to improve the handling, placing, and finishing of the concrete. This is particularly true for self-consolidating concrete and for some pumped concretes. This tends to lead to a higher water demand in the concrete and to higher potential shrinkage. If this shrinkage is not properly taken into account, excessive shrinkage cracking and a consequent loss in durability may occur.

3.3.1.6 Recycled concrete aggregates

There is increasing interest in using recycled concrete as coarse aggregate for new concrete construction. This interest derives from economic considerations and from a desire to make the concrete industry more sustainable. Note that the discussion here is limited to *coarse* aggregate; the fine aggregate resulting from the demolition of building wastes contains some old cement paste and mortar, which will in turn lead to problems with strength, and increases in the drying shrinkage and creep (Alexander and Mindess, 2005). Indeed, there is a RILEM report (Hansen, 1990) that recommends that all material less than 2 mm not be used.

Inevitably, even properly recycled concrete will contain traces of dirt, metals, glass, plastic, and so on. Thus, apart from generally leading to lower strengths in concrete made with recycled concrete aggregate, there is a greater likelihood of durability problems because of these impurities. Such aggregates should be used with caution in concretes that are to be subjected to severe environmental exposures.

3.3.2 Chemical mechanisms

Chemical durability problems may result from interactions between certain types of aggregates and the hydrated cement paste. The most common of these is AAR (see Chapter 9) but adverse reactions involving sulfates also occur. Aggregate sources may also occasionally contain organic or inorganic materials, which affect the setting of the fresh concrete.

Most aggregates, particularly siliceous ones, are immune from acid attack, but calcareous aggregates will be attacked by acids. This does not generally pose a problem, however, and it may be beneficial in helping to reduce the rate or severity of acid attack on the concrete. On the other hand, siliceous aggregates may be susceptible to attack by strong alkalis, as in the AARs discussed in Chapter 9.

However, it should be emphasized that attack by acids, alkalis, or other aggressive chemical agents are much more likely to damage the cementitious phases than the aggregate. Dense, sound aggregates almost always are sufficiently resistant to chemical attack, particularly when embedded in concrete.

3.4 WATER

In general, any water that is drinkable may be used to make concrete. However, some waters that are not drinkable may also be used satisfactorily. Excessive impurities may cause durability problems, and so limits on impurities are given in Table 3.5 (adapted from data in Mindess et al. [2003] and Kosmatka et al. [2002]).

As a further check on the quality of the mixing water, compressive strength of cubes or cylinders tested at 7 days should be at least 90% of the strength of control specimens cast with potable water. In some jurisdictions, such as South Africa, a value of 85% has been adopted.

Seawater contains, on average, about 34,000 ppm dissolved salts, mostly sodium chloride, plus magnesium chloride, magnesium sulfate, calcium sulfate, and calcium chloride. When used to produce plain (unreinforced) concrete, a strength loss of 10–20% can be expected. It should not be used for reinforced concrete, and particularly not for prestressed concrete, as it increases considerably the risk of corrosion of the steel. See Chapter 9 (Section 9.2.6) for a more detailed discussion of the effects of seawater.

Table 3.5 Tolerable levels of impurities in mixing water

Impurity	Maximum concentrations, ppm (mg/L)	Remarks
Dissolved solids	2000	
	1000	If alkali carbonate or bicarbonate is present
Suspended solids (turbidity)	2000	Silt, clay, organic matter
Algae	1000	Entrain air
$Na_2CO_3 + NaHCO_3$	2000	Affect setting
$Ca(HCO_3)_2$, $Mg(HCO_3)_2$	400	Affect setting and may reduce strength
Inorganic acids	10,000	pH not less than 3.0
Sugar	500	Affects setting behavior
Na_2SO_4, $MgCl_2$, $MgSO_4$, $CaCl_2$	10,000	May decrease setting time; reduce ultimate strength
NaCl (in reinforced concrete)	1500	May decrease setting time; may increase the risk of corrosion in reinforced concrete
NaOH	500	May reduce strength
Sodium sulfide	100	
Iron salts	40,000	
Salts of Zn, Ca, Pb, Mn, Sn	500	Retard set
Phosphates, arsenates, borates	500	Retard set
Chlorides (by mass of cement), as per ACI 318 Building Code (2002)	0.06%	Prestressed concrete
	0.15%	Reinforced concrete exposed to chlorides in service
	1.0%	Reinforced concrete kept dry in service
	0.3%	Other reinforced concrete

Source: Adapted from data in Mindess, S., Young, J. F., and Darwin, D., 2003, *Concrete*, 2nd edition, Prentice-Hall, Upper Saddle River, NJ, 644pp; Kosmatka, S. H., Kerkhoff, B., Panarese, W. C., MacLeod, N. F., and McGrath, R. J., 2002, *Design and Control of Concrete Mixtures*, 7th Canadian edition, Cement Association of Canada, Ottawa, Canada, 356pp.

3.5 ADMIXTURES

Most modern concretes contain some combination of both SCMs and chemical admixtures, as described in detail in Aïtcin and Flatt (2016). However, while these materials may not always lead to the desired outcomes, in themselves they do not cause durability problems. An improper choice of admixtures may lead to rheology problems, alterations of the time of set, delays in the rate of strength gain, and so on, but these are not durability problems per se.

3.5.1 Supplementary cementing materials

At one time, the two main SCMs were fly ash (whose use goes back to the 1930s) and GGBFS (whose use in Europe goes back to the early 1900s). Originally, these were used primarily for economic reasons, as they were considerably cheaper than Portland cement. However, it was soon recognized that they could considerably improve the long-term properties of concrete through their pozzolanic reaction with the lime liberated by the hydration of Portland cement, though sometimes at the cost of a rather slower rate of strength gain. Of course, depending upon their source, these materials are sometimes of variable quality and precise chemical composition. However, in and of themselves, they do not cause durability problems for concrete.

Over time, driven by economic considerations (as fly ash and slag became more expensive), a large number of other materials (either naturally occurring or "waste" materials as byproducts of other industries) with pozzolanic properties have come into use. The most common of these are silica fume and metakaolin; various other sources of pozzolanic materials, such as rice husk ash, are also sometimes exploited in specific local markets. While the long-term effects of these materials in concrete are still unknown, there is as yet no evidence to suggest that they themselves lead to durability problems. The same may be said of the much less reactive *filler* materials, such as ground limestone, which are now also commonly added to Portland cement in relatively small quantities (<10% in North America, but up to 20% or even more in Europe).

3.5.2 Chemical admixtures

The picture is somewhat less clear when we look at chemical admixtures. These now include air-entraining admixtures, water reducers, superplasticizers, corrosion inhibitors, rheology modifiers, set-controlling admixtures (retarders or accelerators), water-proofing agents, internal curing materials, and so on. Again, these materials by themselves are not known to create durability problems. However, modern high-performance concretes, which almost always contain a variety of both mineral and chemical admixtures, are becoming increasingly complex. Inevitably, there are chemical interactions among the Portland cement and the various admixtures. This often makes it difficult to find (empirically) the right combination of binder materials and admixtures for any particular project. It is still unclear whether these combinations of materials will be durable in the long run, particularly in very aggressive environments.

3.6 FIBERS

Over the last 50 years, there has been a steady increase in the use of fibers in concrete, for both cast-in-place and shotcrete applications. The fibers are not added to increase strength, though modest improvements in strength may occur. Rather, the fibers are added to help control cracking, and more particularly to impart some postcracking ductility to the concrete (Bentur and Mindess, 2007). A number of different fiber types are currently being used, including steel, polypropylene, polyethylene, carbon, aramid, glass, cellulose, and natural organic fibers such as sisal and jute. (Asbestos fibers were once common, but are not now used because of the health risks associated with the production and handling of these fibers.) Some of the fibers such as carbon, aramid, polyethylene, and polypropylene are completely inert in concrete. Others may react either with the cement or the environment: steel may rust, natural organic fibers may deteriorate over time in damp or alkaline environments, and glass may be attacked by the alkaline environment in concrete.

The fibers themselves do not have any direct effect on the durability of the concrete. However, those fibers that do interact with the surrounding concrete may lose their effectiveness over time.

REFERENCES

Aïtcin, P.-C. and Flatt, R. J. (eds.), 2016, *Science and Technology of Concrete Admixtures*, Woodhead Publishing, Cambridge, UK and Waltham, MA, 613pp.

Aïtcin, P.-C. and Mindess, S., 2015, Back to the future, *Concrete International*, 37(5), 37–40.

Alexander, M. and Mindess, S., 2005, *Aggregates in Concrete*, Taylor & Francis, Abingdon, UK, 435pp.

ASTM C131-14, 2014, *Standard Test Method for Resistance to Degradation of Small-Size Coarse Aggregate by Abrasion and Impact in the Los Angeles Machine*, ASTM International, West Conshohocken, PA.

ASTM C535-12, 2012, *Standard Test Method for Resistance to Degradation of Large-Size Coarse Aggregate by Abrasion and Impact in the Los Angeles Machine*, ASTM International, West Conshohocken, PA.

Bentur, A. and Mindess, S., 2007, *Fibre Reinforced Cementitious Composites*, 2nd edition, Taylor & Francis, Abingdon, UK, 601pp.

Bentz, D. P., Bognacki, C. J., Riding, K. A., and Villareal, V. H., 2011, Hotter cements, cooler concretes, *Concrete International*, 33(1), 41–48.

Blight, G. E. and Alexander, M. G., 2011, *Alkali-Aggregate Reaction and Structural Damage to Concrete*, Taylor & Francis, London, UK, 234pp.

BS EN 197-1:2011, 2011, *Cement Composition, Specifications and Conformity Criteria for Common Cements*, British Standards Institution, London, UK.

Callan, E. J., 1952, Thermal expansion of aggregates and concrete durability, *Journal of the American Concrete Institute, Proceedings*, 48, 504–511.

CSA A23.1-14, 2014, *Concrete Materials and Methods of Concrete Construction*, CSA Group, Toronto, Canada.

de Belie, N., Verselder, H. J., de Blaere, B., van Nieuwenburg, D., and Verschoore, R., 1996, Influence of the cement type on the resistance of concrete to feed acids, *Cement and Concrete Research*, 26, 1717–1725.

Edwards, A. G., 1966, Shrinkage and other properties of concrete made with crushed rock from Scottish sources, *British Granite and Whinstone Federation Journal*, 6(2), 23–42.

EN 197-1, 2011, *Cement—Part 1: Composition, Specifications and Conformity Criteria for Common Cements*, European Committee for Standardization.

Hansen, T. C. (ed.), 1990, *Recycling of Demolished Concrete and Masonry*, RILEM Report No. 6, CRC Press, Boca Raton, FL.

Kim, H.-S., Lee, S.-H., and Moon, H.-Y., 2007, Strength properties and durability aspects of high-strength concrete using Korean Metakaolin, *Construction and Building Materials*, 21, 1229–1237.

Kosmatka, S. H., Kerkhoff, B., Panarese, W. C., MacLeod, N. F., and McGrath, R. J., 2002, *Design and Control of Concrete Mixtures*, 7th Canadian edition, Cement Association of Canada, Ottawa, Canada, 356pp.

Lee, J., Xi, Y., and Willam, K., 2008, Properties of concrete after high-temperature heating and cooling, *ACI Materials Journal*, 105(4), 334–341.

Liu, X., Ye, G., De Schutter, G., Yuan, Y., and Taerwe, L., 2008, On the mechanism of polypropylene fibres in preventing fire spalling in self-compacting and high-performance cement paste, *Cement and Concrete Research*, 38, 487–499.

Mindess, S., and Aïtcin, P.-C., 2014, How can we move from prescription to performance? In H. Bjegovic, H. Beushausen, and M. Serdar (eds.) *Proceedings of the RILEM International Workshop on Performance-Based Specification and Control of Concrete Durability*, RILEM Proceedings PRO 89, RILEM Publications, Paris, pp 267–273.

Mindess, S., Young, J. F., and Darwin, D., 2003, *Concrete*, 2nd edition, Prentice-Hall, Upper Saddle River, NJ, 644pp.

Phan, L. T., Lawson, J. R., and Davis, F. L., 2001, Effects of elevated temperature exposure on heating characteristics, spalling, and residual properties of high performance concrete, *Materials and Structures*, 34, 83–91.

Piasta, J., 1984, Heat deformation of cement paste phases and the microstructure of cement paste, *Materials and Structures*, 17(102), 415–420.

Piasta, J., Sawicz, Z., and Rudzinski, L., 1984, Changes in the structure of hardened cement paste due to high temperature, *Materials and Structures*, 17, 415–420.

Skalny, J., Marchand, J., and Odler, I., 2002, *Sulfate Attack on Concrete*, Spon Press, London, UK, 217pp.

Tamimi, A. K., 1997, High-performance concrete mix for an optimum protection in acidic conditions, *Materials and Structures*, 30, 188–191.

Torii, K. and Kawamura, M., 1994, Effect of fly ash and silica fume on the resistance of mortar to sulphuric acid and sulphate attack, *Cement and Concrete Research*, 24, 361–370.

Chapter 4

Concrete deterioration

4.1 INTRODUCTION

This book is about concrete durability, which is "the ability of a structure or component to withstand the design environment over the design life, without undue loss of serviceability or need for major repair" (ASTM E632, 1996). Durability is a material performance concept and is not an intrinsic material property; a concrete that is durable in one environment may not be durable in another.

Durability is associated with deterioration of concrete over the intended service life of the structure. Figure 4.1 (Owens, 2009) illustrates the progress of deterioration over time of two structures, A and B. Structure A deteriorates at a certain rate but nevertheless accomplishes its expected service life. In contrast, Structure B deteriorates more rapidly, requiring repair during its service life. This deterioration may be expected and even planned for in the case of reliable deterioration models, or it may be unexpected due to inadequate understanding of deterioration mechanisms, or change of external environment or use of the structure. Increasingly, owners are demanding that designers and contractors provide assurance of a predefined, repair-free service life of concrete structures.

Deterioration of concrete is an extensive topic, and this chapter cannot give a comprehensive treatment (more information can be obtained from the references given in "Further Reading" and "References"). It is the subject of extensive ongoing investigations, and is thus a developing science and much still needs to be understood. Nevertheless, a basic understanding of concrete deterioration is necessary in order to grasp and apply the durability principles laid out in other chapters of this book. This chapter lays out the framework for understanding deterioration and deals with important concrete deterioration mechanisms. It draws largely on material from a chapter on "Durability of Concrete" in Owens (2009).

Factors affecting deterioration of concrete are summarized in Figure 4.2 (Owens, 2009). Deterioration results from interactions between the concrete system, specifically the ability of the concrete to resist deterioration, and its environment or the degree of aggressiveness that the concrete must

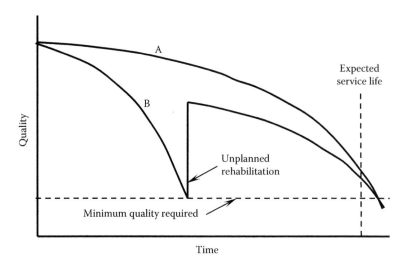

Figure 4.1 Two scenarios of deterioration and rehabilitation of a structure. (Reproduced from Owens, G. (ed.), 2009, *Fulton's Concrete Technology*, Cement and Concrete Institute, Midrand, South Africa, Figure 9.1, with permission from the Concrete Institute.)

withstand. Not all the factors in Figure 4.2 may be important all the time. However, critical factors are binder type and water/binder (w/b) ratio (intrinsic factors), and curing and early-age temperature history (extrinsic factors), while the most critical environmental factor will be aggressiveness of agents of attack. Cracking of concrete is also critical to its deterioration, and usually cracking from any cause will accelerate concrete deterioration by allowing further ingress of aggressive agents. Cracking is often also a consequence of deterioration, thus setting up an "accelerating cycle."

A distinction must be made between unreinforced and reinforced concrete. The former is often used in large masses such as dams and filling of excavations, and deterioration mechanisms relate to the concrete fabric itself. In reinforced concrete, there is the additional issue of corrosion of embedded reinforcing steel. Since this is such a pervasive form of deterioration, it is covered in a separate section.

Premature deterioration of concrete structures has severe consequences, such as threat to economic growth, depletion of natural and nonrenewable resources, impacts on human safety, and costs (Gjørv, 2009). The economic losses alone are substantial with the annual cost of corrosion worldwide estimated at US$ 2.2 trillion (2010), which is about 3% of the world's gross domestic product (GDP) of US$ 73.33 trillion (Al Hashem, 2011), and concrete corrosion contributes to this. Taking the example of the Arabian Gulf, the annual cost of repair and rehabilitation due to corrosion in the United Arab Emirates (UAE) is estimated at US$ 14.26 billion (2011), which is

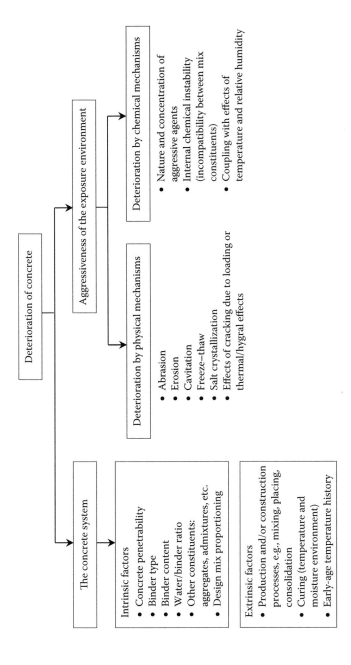

Figure 4.2 Factors influencing the deterioration of concrete. (Reproduced from Owens, G. (ed.), 2009, *Fulton's Concrete Technology*, Cement and Concrete Institute, Midrand, South Africa, Figure 9.2, with permission from the Concrete Institute.)

about 5.2% of the country's GDP over 3 years (2009–2011) (Al Hashem, 2011). These examples of costs of repair and maintenance of reinforced concrete structures serve to illustrate the scale of the problem and to highlight the threat to the concrete construction industry worldwide.

4.2 CONCRETE PROPERTIES RELEVANT TO DETERIORATION

Concrete deterioration is largely due to the ingress of aggressive agents from the environment into the matrix. Other deterioration may arise internally, for example, alkali–aggregate reaction (AAR) or delayed ettringite formation. Even internal causes of deterioration require movement of ions or liquids, so that deterioration is closely related to the concrete's transport properties, which also include electrical transfer since steel corrosion requires internal movement of ions to carry electrical current.

4.2.1 Transport properties of concrete

Deterioration mechanisms such as chemical attack, leaching, chloride ingress, or carbonation relate to the ease with which fluids or ions can move through the concrete microstructure. The term used to describe this is *penetrability*, which is the degree to which concrete permits gases, liquids, or ionic species to move through its pore structure; it includes permeation, sorption, diffusion, and migration. A representation of the various penetrability processes in a typical sea wall structure (excluding migration) is given in Figure 4.3 (after BS 6349-1), while Figure 4.4 shows interactions of these processes (Claisse, 2005).

4.2.1.1 Permeation

Permeation is the movement of fluids through the pore structure under a pressure gradient, with the pores saturated with the particular fluid. It is expressed as a permeability coefficient, which quantifies the fluid-transfer capacity of the concrete by permeation. Concrete permeability depends on its microstructure, its moisture condition, and the characteristics of the permeating fluid. Regarding deterioration, permeation is most detrimental where the permeating fluid is aggressive, for example, soft water permeating through a tunnel lining and leaching soluble components of the concrete thereby weakening it.

4.2.1.2 Absorption

Absorption occurs when a fluid is drawn into a porous unsaturated material due to capillary suction, and depends on the pore geometry and the

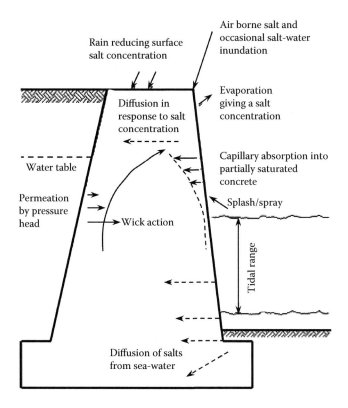

Rain reducing surface salt concentration

Air borne salt and occasional salt-water inundation

Diffusion in response to salt concentration

Evaporation giving a salt concentration

Water table

Capillary absorption into partially saturated concrete

Permeation by pressure head

Splash/spray

Wick action

Tidal range

Diffusion of salts from sea-water

Figure 4.3 Penetrability processes illustrated by way of a sea wall. (After BS 6349-1-4:2013, 2013, Maritime works. General. Code of practice for materials, The British Standards Institution, London, UK.)

degree of saturation. It is usually of concern in wetting and drying cycles at the concrete surface, which can result in softening, cracking, and the build-up of salts. Absorption is strongly influenced by the size and interconnectedness of the larger capillaries and is sensitive to curing, particularly of the concrete surface.

4.2.1.3 Diffusion

Diffusion is the process by which liquids, gases, or ions are transported through a porous material by a concentration gradient. Diffusion occurs in partially or fully saturated concrete and is an important transport mechanism for many concrete structures exposed to salts, which diffuse inwards due to a high surface concentration. The material property for diffusion is diffusibility, measured typically by a diffusion coefficient.

Diffusion is described by Fick's first and second laws of diffusion. In concrete, diffusion is complicated by several factors: chemical

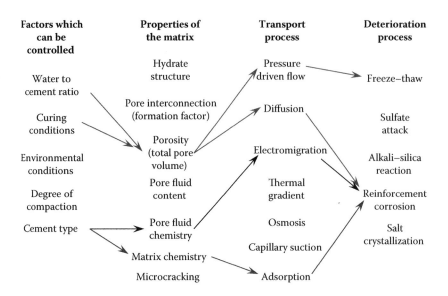

Figure 4.4 Concrete and environment: factors influencing "penetrability" of concrete. (After Claisse, P. A. 2005. Transport properties of concrete, *Concrete International*, 27(1), 43–48, reproduced with permission from the American Concrete Institute.)

interactions with the hydration products, partially saturated conditions, and defects such as cracks and voids, all of which render the assumptions behind Fick's laws largely invalid. Nevertheless, these laws are still used by resorting to "apparent" or "effective" diffusion coefficients for analysis and prediction.

In maritime structures or structures exposed to deicing salts, chloride diffusion may cause depassivation of the reinforcing steel, leading to destructive corrosion. (See later in this chapter on steel corrosion.) The transport of oxygen in concrete to the steel surface—necessary for active corrosion—is also governed mainly by diffusion.

4.2.1.4 Migration

Migration (or accelerated diffusion, electrodiffusion, or conduction) is the movement of ions in a solution under an electrical field, typically used in accelerated chloride tests. It is described by the Nernst–Planck equation. Details may be found in Andrade (1993) and in Chapter 6.

4.2.1.5 Transport properties of cracked concrete

The presence of cracks substantially modifies concrete's transport properties and therefore also deterioration mechanisms and rates. Discrete and well-distributed microcracks will influence transport in a very different manner compared to visible, connected, localized macrocracks. Transport in cracked concrete is a coupled phenomenon between the matrix and the crack. The parameters for describing flow in cracked versus intact material are different; crack properties dominate for cracked concrete, while permeability and porosity govern for sound material.

Theories and processes that hold in sound concrete cannot therefore simply be transferred to cracked concrete. The cracks themselves are the most important element, represented by crack width and shape, crack density/frequency, degree of crack connectivity, and crack origin. (Further information is given in Section 4.2.3.)

4.2.1.6 Summary of transport processes

Transport processes seldom act purely in isolation, although frequently one or other process dominates. When considering concrete deterioration, the dominant as well as the secondary transport processes must be identified for a proper diagnosis. This is also critical for durability design, in which models or procedures are selected based on the transport processes in play.

4.2.2 Mechanical, physical, and chemical properties of concrete in relation to deterioration

Deterioration involves interactions between the material and the environment, with transport properties, as well as basic mechanical, physical, and chemical properties of concrete, playing crucial roles.

4.2.2.1 Mechanical and physical properties

Structural engineers generally understand the mechanical (and to a lesser extent, physical) properties of concrete, since they are used in design and specification. An understanding of these properties is also essential to understanding concrete deterioration.

The chief mechanical and physical properties of concrete are compressive and tensile strength (and related shear strength), stiffness (elastic modulus), creep and shrinkage, coefficient of thermal expansion, and density. Of these, strength and stiffness predominate in structural design. Strength has been used traditionally as a proxy for durability and resistance to deterioration. However, with developments in concrete technology and practice, this approach is now wholly inadequate and overly simplistic. Modern binders are highly variable, new technologies such as self-compacting concrete are becoming common-place, and admixtures permit mix proportions to be adjusted within wide ranges, thus rendering the "strength–durability equivalence" of the past obsolete. This is illustrated in Figure 4.5 with data from a site-based study in which reinforced concrete structures were assessed for

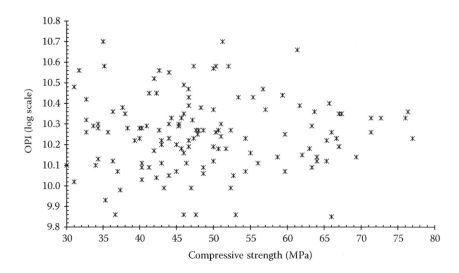

Figure 4.5 Scatter plot of OPI for concrete structures assessed *in situ* versus standard cube strength. (Reprinted from Construction and Building Materials, 45, Nganga, G., Alexander, M. G., and Beushausen, H., Practical implementation of the durability index performance-based design approach, 251–261, Copyright 2013, with permission from Elsevier.)

strength based on standard cube testing, and *in situ* permeability using an oxygen permeability index (OPI) test. Permeability and strength essentially show no relationship, and strength clearly cannot be used as a proxy for "durability" in this case. Similar situations arise for other properties such as diffusibility, particularly with modern blended binders where the binder chemistry is equal to, if not more important than, the w/b ratio or strength.

Stiffness is a mechanical property not always appreciated for the profound effect it can have on deterioration, more particularly cracking. In situations where concrete is restrained against movement, leading to induced shrinkage and/or thermal strains, tensile stresses are set up that can crack the concrete. For any given restraint, the corresponding tensile stress is directly proportional to the elastic modulus of the concrete. Thus, high-modulus concretes are more crack-sensitive and therefore more durability-sensitive. These effects can often be observed in slabs on grade or concrete pavements, in wall sections cast on heavy bases, in changes in section shape such as openings, and the like. Typical photographic illustrations are shown in Figure 4.6.

4.2.2.2 Chemical properties

The chemical products of hydration are stable in many environments, but they can be attacked, for example, by sulfates or soft water. Also, the

(a)   (b)

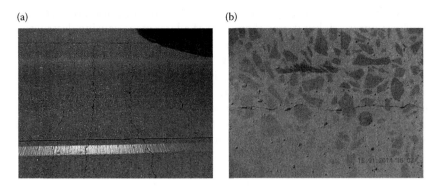

Figure 4.6 Photos of concrete cracking due to restrained shrinkage or thermal movement. (a) Cracking of concrete slab-on-grade . (Courtesy of Dr P. Strauss); (b) thermal cracking of concrete wall (sawn section).

hydrates often interact with penetrating agents as in carbonation or chloride ingress, forming new compounds which may or may not be detrimental to the concrete.

The main products of hydration when Portland cement reacts with water at normal temperatures are the following (approximate compositions):

Silicate phases	$3CaO \cdot 2SiO_2 \cdot 8H_2O$—calcium silicate hydrates [CSHs]
Aluminate phases	$3CaO \cdot Al_2O_3 \cdot 3CaSO_4 \cdot 32H_2O$—ettringite or AFt, formed by reaction of tricalcium aluminate (C_3A) with gypsum and water
	$3CaO \cdot Al_2O_3 \cdot CaSO_4 \cdot 12H_2O$—monosulfate or AFm, formed by transformation of ettringite when sulfate is exhausted

The silicate reaction products are poorly crystalline and colloidal, with a byproduct of crystalline calcium hydroxide in fairly abundant quantities. The silicate hydrates are stable in high pH environments such as in concrete. However, percolating waters or surface leaching can remove more soluble phases such as calcium hydroxide and metal hydroxides of sodium and potassium, thus destabilizing the CSH and leading to progressive breakdown of the matrix as additional calcium hydroxide is liberated to reestablish equilibrium.

The aluminate phases are thermodynamically less stable than the silicate phases, and their reactions can move in either direction depending on the concentration of soluble sulfates present. This can occur internally in the matrix, or by external ingress of sulfate solutions, which then favors formation of expansive ettringite.

The hydration reactions are exothermic, giving rise to heat gain and temperature rise in hydrating concrete at early ages. If the temperature rises to excessively high levels (typically >70–80°C), the formation of ettringite may be suppressed, but it may then form at a later stage causing destructive expansion (so-called delayed ettringite formation [DEF]).

In order to ensure the longevity of concrete structures, its inherent chemical stability together with its physical robustness are critical. Supplementary cementitious materials (SCMs), which increase the quantity of hydrates and also generally improve the concrete microstructure, have also made a large contribution to improved resistance of concrete to deterioration. In aggressive environments, blended cements are almost always preferred, with the important proviso that blended cement concretes are usually more sensitive to poor curing and low temperatures, and suitable precautions must be taken to address these issues during construction.

4.2.3 Cracking of concrete and its influence on deterioration

Cracking is ubiquitous in concrete, and almost all concrete structures exhibit cracking of some kind, making cracking a deterioration mechanism in its own right. Cracking can occur from the microscale, where differences exist between thermal and hygral coefficients of the constituents of concrete (e.g., within the hydrated cement paste [hcp] or between hcp and the aggregates), to the macroscale such as large cracks several millimeters wide from excessive load, drying shrinkage cracking, or AAR. Micro-versus macrocracking will have different effects on concrete properties and deterioration mechanisms.

Cracking can be divided broadly into structural cracking and non-structural (or non-load-induced) cracking. Structural cracking is often the intent of design, such as cracked beams that function by mobilizing the tensile properties of the reinforcing steel. Design formulae can aid in estimating and controlling the extent of structural cracking, and this cracking will not be further considered here. Suffice to say that structural cracking has a profound influence on the deterioration of concrete, such as its influence on corrosion rates of reinforced concrete elements in aggressive environments where active corrosion may commence immediately the structure is put into service, thus potentially reducing the service life.

Non-structural cracking usually involves interaction of the concrete and its environment, and may seriously affect concrete durability by allowing ingress of aggressive agents into the body of the material. It is often related to restrained volume changes in the constituent materials. In general, unrestrained concrete does not crack, or at least does not show macrocracking. Restraint applies at different levels of scale: at a microscale, hcp undergoes temperature and moisture movements that are restrained by the stiff aggregates causing internal microcracking; at a macroscale, slabs-on-grade are restrained by the sub-base and may crack in the absence of properly designed and executed joints (Figure 4.6). Thus, preventing or minimizing cracking involves reducing restraint in concrete.

In the rest of this section, cracking mechanisms and causes related to the various types of cracks, mainly non-structural, are reviewed with an emphasis on concrete deterioration. Figure 4.7 summarizes the main types of non-structural cracking together with causes and informative comments, while Figure 4.8 illustrates some of these cracks.

Figure 4.7 shows that cracking can occur in both the plastic state and the hardened state. Plastic-state cracking tends to be of the macro type with serious consequences for structural integrity and durability. Some plastic cracking can be remedied during construction by re-vibrating the concrete provided it is still sufficiently plastic. However, cracking in the hardened state results in permanent damage to the concrete with attendant consequences.

Figure 4.7 also suggests the likelihood of which type of cracking will be micro- versus macroscale. The distinction between these two scales of cracking is somewhat arbitrary, and indeed some researchers insert a further scale—the mesoscale. Here, microcracking is taken as of the order of cement grain size, that is, up to about 30 microns, in general not visible to the naked eye. Macrocracking is taken to be in the order of 100 microns up to several millimeters, and will generally be visible. Mesoscale falls between the other two categories.

Microcracking will compromise the transport properties of concrete by increasing its permeability and penetrability. However, this will depend on the microcrack width, and the concrete may continue to act largely as a "continuum." Figure 4.9 illustrates the effect of microcracking on water permeability through concrete laboratory specimens. Above a crack width of about 100 microns, the effects become magnified.

Macrocracking by contrast represents major discontinuities in the material, with orders of magnitude increases in the penetrability, often leading to substantial reduction in the service life of a structure. In Figure 4.10, the macrocrack from AAR will allow rapid ingress of aggressive environmental agents to the steel, leading possibly to additional deterioration by way of steel corrosion and further corrosion-induced cracking. Figure 4.11 shows the effects of cracked versus uncracked concrete on rates of steel corrosion in small concrete beams, where it is evident that even relatively "small" macrocracks can have substantial effects on this phenomenon.

While cracking can be considered a deterioration mechanism on its own, it usually interacts with other mechanisms of deterioration, either causing an increase in the rate of deterioration or appearing as a consequence. Thus, cracking can cause steel corrosion to initiate, but this corrosion can induce additional cracking which further exacerbates the corrosion, and so on. Cracking and its influence on concrete deterioration

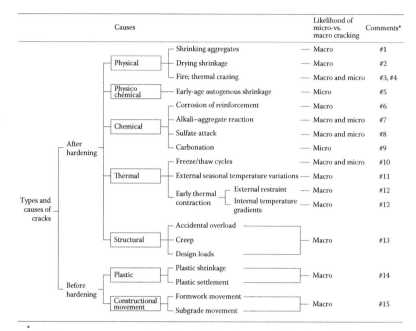

#1 Shrinking aggregates can cause excessive cracking and deflections in structures. Eliminate by standard dimensional stability testing.

#2 Drying shrinkage caused by moisture loss from the concrete. Most likely in thin elements that dry rapidly.

#3 Concrete subjected to fire loses moisture in the outer layer and also suffers de-hydration and destructive reactions, giving both macro and micro-cracking.

#4 Crazing is evident on "fair-face" concrete with impermeable formwork, or where slabs are over-trowelled bringing fines to the surface; fine "map cracking".

#5 Autogenous (internal) shrinkage results from both "chemical shrinkage" and self-desiccation shrinkage, particularly in rich mixtures with low w/b ratios (<0.4 typically). It can cause extensive internal damage.

#6 See other section in this chapter for cracking and reinforcement corrosion.

#7 AAR cracking, normally seen as coarse "map cracking", is both macro and micro scale, resulting from expansive reactions between usually cement alkalis and susceptible forms of reactive aggregate minerals, typically reactive silica. It is highly detrimental to aesthetics and also to structural integrity depending on the nature and form of the structure.

#8 Sulphate attack arises from volume expansion of the cementitious products, and is generally highly damaging to concrete.

#9 Carbonation results in shrinkage of the matrix by release of water that then evaporates. A secondary form of shrinkage.

#10 Freeze-thaw damage produces internal microcracks that also manifest as macrocracks on the surface. This cause also includes aggregate D-cracking in which aggregates crack from internal hydraulic pressure on freezing.

#11 Temperature variations can cause surface cracking which, though visible, may not penetrate very deeply.

#12 Early thermal contraction arises from concrete temperature rise due to heat of hydration, and is often exacerbated by subsequent drying shrinkage. Occurs typically in restrained elements such as walls cast on stable bases.

#13 Structural-induced cracking is of many types, often by design (e.g. in flexural members). Where it is non-intentional, it is usually detrimental to durability.

#14 Plastic shrinkage cracking arises mainly in flat work where the rate of evaporation from the surface exceeds the rate of belling, normally associated with high temperatures and drying winds; plastic settlement cracking is caused by settlement of the fresh concrete over reinforcement, or at changes of depth in e.g. troughs and waffle slabs.

#15 Formwork or subgrade movement while the concrete is still plastic can also cause macrocracking.

Figure 4.7 Types and causes of cracks in concrete. (Adapted from Concrete Society TR22, 2010, *Non-Structural Cracks in Concrete*, 4th edition, Concrete Society, Surrey, UK.)

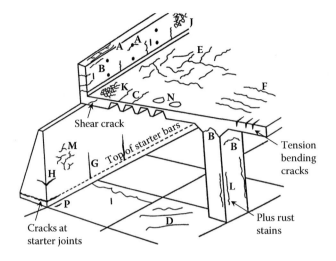

Shear crack

Top of starter bars

Tension bending cracks

Cracks at starter joints

Plus rust stains

Legend:

Type of cracking	References	Subdivision	Most common locations	Primary causes (excluding restraint)	Secondary causes	Remedy	Time of appearance
Plastic settlement	A	Over reinforcement	Deep sections	Excess bleeding	Rapid early drying conditions	Reduce bleeding (air entrainment) or revibrate	10 minutes to 3 hours
	B	Arching	Top of columns				
	C	Change of depth	Trough and waffle slabs				
Plastic shrinkage	D	Diagonal	Roads and slabs	Rapid early drying	Low rate of bleeding	Improve early curing	30 minutes to 6 hours
	E	Random	Reinforced concrete slabs				
	F	Over reinforcement	Reinforced concrete slabs	Ditto + steel bar near surface			
Early thermal contraction	G	External restraint	Thick walls	Excess heat generation	Rapid cooling	Reduce heat and/or insulate	1 day to 3 weeks
	H	Internal restraint	Thick slabs	Excess temp. gradients			
Long-term drying shrinkage	I	–	Thins slabs/walls	Inefficient joints	Excess shrinkage, inefficient curing	Reduce water content, improve curing	Several weeks or months
Crazing	J	Against formwork	"Fair-faced" concrete	Impermeable formwork	Rich mixes, poor curing	Improve curing and finishing	1–7 days, sometimes much later
	K	Floated concrete	Slabs	Over trowelling			
Corrosion of reinforcement	L	Natural	Columns and beams	Insufficient cover	Poor quality concrete	Eliminate causes listed	More than 2 years
Alkali–silica reaction	M	–	(Damp locations)	Reactive aggregates + high alkali content		Eliminate causes listed	More than 5 years

Figure 4.8 Illustration of non-structural cracks in concrete. (Adapted from Concrete Society TR22, 2010, *Non-Structural Cracks in Concrete*, 4th edition, Concrete Society, Surrey, UK.)

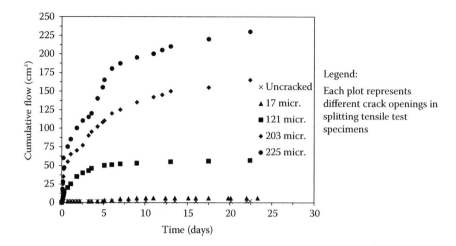

Figure 4.9 Influence of microcracking on normal strength concrete permeability (measured as cumulative flow). (With kind permission from Springer Science+Business Media: Materials & Structures, Permeability of cracked concrete, 32, 1999, 370–376, Aldea, C. M., Shah, S. P., and Karr, A.)

is a complex subject which deserves fuller treatment than can be given here. (Additional information can be obtained from the references given in "Further Reading.")

4.3 MECHANISMS OF DETERIORATION*

4.3.1 Introduction

For the sake of classification, deterioration mechanisms can be divided into two broad categories: those caused mainly by physicomechanical effects and those caused mainly by chemical effects (with biodegradation mechanisms as a subset of chemical deterioration mechanisms). These categories will be discussed in the following sections using the summary given in Table 4.1, which outlines key aspects. Detailed information on deterioration mechanisms should be sought in specialized texts (Biczók, 1964; Heckroodt, 2002; Soutsos, 2010; Alexander et al., 2013). Since corrosion of steel reinforcement in concrete is such an important and pervasive form of deterioration, this receives special attention later.

Since concrete deterioration is dependent on interactions between the concrete and its environment, the probability of occurrence and the rate of deterioration will depend on the nature and aggressiveness of the

---

* A detailed account of concrete deterioration mechanisms is beyond the scope of this chapter. See specialist texts in "Further Reading."

Figure 4.10 AAR-induced cracking representing a macrocrack, which may allow further deterioration to occur (core diameter 100 mm).

environment as well as the physical and chemical resistance of the concrete. General principles are as follows:

- Deterioration usually involves the movement of liquids, gases, and ions through the concrete pore structure, hence the importance of transport properties of concrete.
- Water in the environment and in the concrete pores is necessary for most forms of deterioration. Dry concrete will generally deteriorate at a negligible rate unless it is subject to, for example, dry abrasion. The presence or absence of water is often crucial in terms of deterioration mechanisms.
- The resistance of concrete to deterioration is very dependent on the binder system used. Blended binders are generally more resistant to

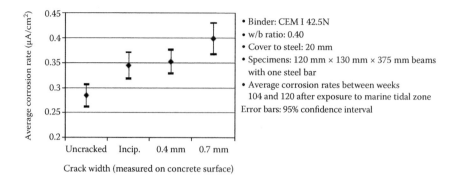

Figure 4.11 Corrosion rate of laboratory beams as a function of surface crack width. *Note:* "Incipient" crack invisible to the naked eye, obtained by loading to the point of nonlinearity of the load–deflection curve. (Courtesy of Otieno, M., 2014, *The development of empirical chloride-induced corrosion rate prediction models for cracked and uncracked steel reinforced concrete structures in the marine tidal zone*, PhD dissertation, University of Cape Town, Cape Town, South Africa.)

many forms of deterioration than plain Portland cement, but each case needs to be assessed.

- Good physical resistance requires concrete that is well-compacted, dense, and well-cured, particularly in the outer layer. In many cases, a durable outer "skin" of concrete is the only practical means of ensuring longevity to a concrete structure.

Regarding the last bullet point, the resistance of specifically reinforced concrete to deterioration relates to the quality of the cover layer. Therefore, it is necessary to distinguish between bulk properties of concrete such as compressive strength or bulk absorption, and properties relating to penetrability that refer more to the surface layer. When deterioration is caused by external aggressive agents, the quality and properties of the surface zone are critically important.

The main mechanisms of concrete deterioration are summarized in Table 4.1, and include the following:

- *Mechanical and physical mechanisms*: Abrasion, erosion, cavitation, freeze–thaw, salt crystallization, cracking due to loading or thermal/hygral effects (e.g., temperature or moisture cycling), and fire damage.
- *Chemical mechanisms*: Atmospheric gases, alkalis, soft and pure waters, waters that are highly mineralized or contain other deleterious substances, acids, sulfates (external and internal), AAR. (Biodeterioration mechanisms, giving rise frequently to acid attack, are considered a subset of chemical mechanisms.)

Table 4.1 Summary of mechanical, physical, chemical, and biodeterioration mechanisms for concrete

Deterioration mechanism	Brief description and primary effect of deterioration	Nature of aggressive medium	Typical examples; likely occurrence and frequency	Avoidance or special precautions
Mechanical and physical deterioration mechanisms				
Abrasion	Wear of the concrete surface by traffic leading to progressive mass loss and reduction in dimensions (incl. cover to reinforcement)	Abrasive traffic, wheeled, tracked, or foot	Wear of concrete slabs and pavements by vehicular traffic; of concrete paving stones by foot traffic. Industrial floors, hard standings, footpaths, container storage areas	Avoid laitance. Well-compacted and cured concrete with good surface finish
Erosion	Wear of the concrete surface by action of fluids containing solid particles in suspension, leading to progressive mass loss and reduction in dimensions	Suspended abrasive solid particles in fluid, typically water	Wear of concrete tunnels, passages, channels and conduits. Water-retaining or -conveying structures such as spillways, canal-linings, etc.	Avoid laitance. Well-compacted and cured concrete with good surface finish
Cavitation	Wear of the concrete surface by action of shock waves produced by collapsing vapor bubbles in liquids, leading to progressive mass loss and reduction in dimensions. Localized pressures as high as 700 MPa can be generated, which will damage even the strongest concretes	High-velocity liquids undergoing sudden change in pressure or direction of flow	Wear of concrete spillways and steep channels. Water-retaining or -conveying structures such as spillways, chutes, etc.	Hydraulic design considerations—limit surface misalignments, avoid abrupt change of slope/direction of flow. Low w/c concrete with smooth surfaces

(*Continued*)

Table 4.1 (Continued) Summary of mechanical, physical, chemical, and biodeterioration mechanisms for concrete

Deterioration mechanism	Brief description and primary effect of deterioration	Nature of aggressive medium	Typical examples; likely occurrence and frequency	Avoidance or special precautions
Freeze–thaw	Expansion of moisture within the concrete matrix (and some coarse aggregates) due to freezing. If expansion exceeds the available pore volume, the concrete is damaged. Repeated freeze–thaw cycles lead to cracking and spalling due to progressive expansion	Water in the pores of concrete matrix in cold climates	Cracking and spalling of exposed elements from progressive expansion of cement paste due to freeze–thaw cycles. Scaling of concrete slabs subjected to freeze–thaw cycles. Cold-climate regions	Well-compacted and cured, low w/b concrete; use of sufficient quantity and spacing of entrained air
Salt crystallization/exfoliation	Recrystallization of salt due to highly concentrated salt solution in the concrete, typically as moisture content decreases. Tensile stresses develop in the cement paste matrix leading to cracking/exfoliation and surface scaling	Highly concentrated salt solutions	Exfoliation of concrete seawalls or slabs subject to salts, industrial plants with salts. Coastal regions, industrial plants, salting of roads	Well-compacted and cured, low w/b concrete
Cracking due to loading or thermal/hygral effects (temperature or moisture cycling)	Cracking caused by cyclic expansion and contraction due to temperature changes	Significant seasonal temperature differential	Cracking of concrete pavements with surfaces exposed to wide range of temperatures while the base is relatively protected. Regions with high-temperature ranges, e.g., arid regions; metallurgical plants with furnaces	Use of concrete with lower coefficient of thermal expansion or low stiffness

(Continued)

Table 4.1 (Continued) Summary of mechanical, physical, chemical, and biodeterioration mechanisms for concrete

Deterioration mechanism	Brief description and primary effect of deterioration	Nature of aggressive medium	Typical examples; likely occurrence and frequency	Avoidance or special precautions
Fire damage	Cracking and (explosive) spalling due to expulsion of free water and vapor at high temperatures	Persistent high-intensity fire	Explosive spalling of concrete members subjected to prolonged high-intensity fire. Statistically rare—usually linked to accidental occurrences	Well-compacted and cured, low w/b concrete made with carbonate aggregates

Chemical deterioration mechanisms (typical governing equations are given at the end of the table)

Exchange reactions between external aggressive medium and components of hardened cement paste (ion exchange or substitution)

Atmospheric gases: carbonation and other acidification	Diffusion of gases such as CO_2 in pore fluid of concrete, and subsequent reaction with calcium compounds to form carbonates, leading to depassivation and corrosion of reinforcing steel	Atmospheric CO_2 in semi-humid environment	Cracking of reinforced concrete elements due to corrosion-induced expansion. Industrial regions or environments with elevated CO_2 in regions with average RH of about 65%	Well-compacted and cured, low w/b concrete; blended cements may perform less well
Alkalis (i.e., hydroxides of sodium or potassium)	Soluble alkalis (sodium and potassium) react with cement paste, affecting solubility of hydrates. They may react to form sodium-containing varieties of AFm (monosulfate). Precipitation of secondary gypsum leads to an increase in volume and expansive stresses. May also attack alkali-susceptible aggregates, e.g., some silicates	High-solubility alkali ions of sufficient concentration (sodium > 25%, potassium > 25%)	Attack of concrete elements frequently exposed to highly soluble alkali-bearing salts. Industrial processes with highly concentrated sodium and/or potassium solutions or salts	Well-compacted and cured, low w/b concrete. Use of calcareous aggregates lessens attack

(Continued)

Table 4.1 (Continued) Summary of mechanical, physical, chemical, and biodeterioration mechanisms for concrete

Deterioration mechanism	Brief description and primary effect of deterioration	Nature of aggressive medium	Typical examples; likely occurrence and frequency	Avoidance or special precautions
Hydrolysis and dissolution of the products of cement hydration (ion removal)				
Soft and pure waters— leaching of concrete	Leaching involving two phenomena: 1. Diffusion of ionic species (Ca^+ and OH^- mainly) in the pores of the cementitious matrix due to concentration gradients between the highly alkaline pore solution and the ion-hungry external solution 2. Dissolution of the Ca^+-bearing cementitious phases by hydrolysis to supply Ca^+ ions to maintain equilibrium	Pure or soft waters very low in dissolved salts; exacerbated by dissolved CO_2 content	Attack of concrete water-conveying or -retaining structures such as pipes, tunnels, channels, dam walls and spillways, etc. Elements exposed to subterranean waters, or waters deriving from hard stable quartz-bearing strata, e.g., quartzitic sandstones	Well-compacted and cured, low w/b concrete
Highly mineralized waters, e.g., brackish waters, magnesium attack in sea water	The products of magnesium (sulfate) attack on cement hydrates include gypsum, brucite, and magnesium silicate hydrates. Brucite causes C–S–H to release lime in order to maintain equilibrium with the pore solution leading to a progressive decalcification of the cement paste	Dissolved salts such as magnesium in seawater, industrial processes with magnesium solutions	Attack of concrete elements in marine environments or certain industrial plants. Coastal or industrial regions	Well-compacted and cured, low w/b concrete

(Continued)

Table 4.1 (Continued) Summary of mechanical, physical, chemical, and biodeterioration mechanisms for concrete

Deterioration mechanism	Brief description and primary effect of deterioration	Nature of aggressive medium	Typical examples; likely occurrence and frequency	Avoidance or special precautions
Acids: mineral and organic	Solubility of the calcium salts in cement systems leads to leaching of cementitious materials, responsible for increased porosity and loss of strength. The dissolution of the hydrates is the consequence of acid–base reactions	Strong acids with soluble salts, e.g., HNO_3 and HCl; acids forming insoluble salts, e.g., oxalates, are less aggressive	Attack of concrete elements frequently exposed to strong acids. Industrial plants with acids; agroprocessing industries	Well-compacted and cured, low w/b concrete

Expansive stresses caused by conversion of the products of hydration by aggressive agents (ion addition) or internal deleterious reactions

Sulfates

External sulfate attack—general	Can take different forms, depending on the particular sulfate (i.e., cation): Cracking and spalling of matrix due to the formation of expansive products, e.g., gypsum and ettringite Decalcification of CSH and breakdown of concrete microstructure Effect of cation: ammonium and magnesium sulfate very aggressive—also participate in reactions with $Ca(OH)_2$. For example, for magnesium sulfate, the expansive reaction causes $Mg(OH)_2$ (brucite) precipitation and CSH decomposition to release $Ca(OH)_2$	Soluble sulfates present in groundwater or soils Industrially derived sulfates (e.g., effluents) from chemical processes Nature of cation important	Attack of concrete foundations by sulfate-rich groundwater Sea water attack by sulfates (mainly magnesium)— ameliorated by chlorides and brucite layer formed on concrete Arid regions with high soil sulfate contents. Industrial processes with sulfates Mining regions Run-off containing fertilizers In descending order of aggressiveness (and hence rate): ammonium, magnesium, sodium/potassium, calcium sulfate	Sulfate-resisting cement or cement extenders (specifically high-slag concretes). Main precaution is dense well-compacted concrete, well-cured, low w/b ratio

(Continued)

Table 4.1 (Continued) Summary of mechanical, physical, chemical, and biodeterioration mechanisms for concrete

Deterioration mechanism	Brief description and primary effect of deterioration	Nature of aggressive medium	Typical examples; likely occurrence and frequency	Avoidance or special precautions
Thaumasite formation	Formed by reaction of calcium silicate hydrates with environmental solutions of sulfates and hydrated carbonates (CO_3^{2-} ions), under moist conditions and normally in cold to cool temperatures (typically 0–10°C) Leads to formation of a soft, white, and pulpy mass and eventual total disintegration of the hydrated cement paste matrix	Soluble sulfates (and ettringite) with carbonate source, e.g., calcareous aggregate, under generally cool to cold temperatures	Attack of moist buried elements, e.g., footings and foundations, with soluble sulfates and typically carbonate aggregates. Concrete elements in cool to cold environments (below 15°C) and calcareous aggregates	Limit the quantity of calcareous aggregate; use of cement extenders such as fly ash; well-compacted and cured, low w/b concrete
Sulfates: internal sulfate attack DEF (internal)	Caused by re-formation of ettringite under normal service conditions, after high-temperature (>65°C) early curing (ettringite unstable above 65°C)	High-temperature curing, particularly with high sulfate content cement	Typically precast elements with high-temperature heat curing are more susceptible, e.g., railway sleepers; depends on curing temperature	Avoid excessive curing temperatures; control sulfate content of cement
AAR (ASR) (internal deleterious reaction)	Reaction between certain alkali-susceptible siliceous constituents in the aggregate and alkali hydroxides in the pores of hydrated cement paste matrix. Leads to the formation of alkali–silica gel that can absorb water resulting in internal stresses, which frequently causes cracking	Alkali–silica gel in cement paste matrix, usually with source of moisture	Concrete elements with reactive aggregate and with sufficient alkalis; exacerbated by frequent exposure to wetting cycles (e.g., rain). Regions with alkali-reactive aggregates and high cement alkalis	Limit the reactive forms of silica in aggregate; low-alkali cements; well-compacted and cured, low w/b concrete

(Continued)

Table 4.1 (Continued) Summary of mechanical, physical, chemical, and biodeterioration mechanisms for concrete

Deterioration mechanism	Brief description and primary effect of deterioration	Nature of aggressive medium	Typical examples; likely occurrence and frequency	Avoidance or special precautions
Biodegradation mechanisms (typical governing equations are given at the end of the table)				
Diffusion of gases, deposition of excreta on bioreceptive surfaces, and dissolution of soluble calcium complexes	Microbially-induced (biogenic) processes that may facilitate: 1. Diffusion of gases such as CO_2 (forms H_2CO_3 in moist environments, which reacts with the calcium in HCP to form carbonates); SO_2 (forms H_2SO_3, and later H_2SO_4, which reacts with the CH in HCP to form expansive gypsum) 2. Excretion of H_2SO_4 by sulfide-oxidizing bacteria—see (1) above for reaction mechanism	Acids with soluble salts	Attack of concrete water-conveying or -retaining structures such as pipes, wet wells, tunnels, etc.	Well-compacted and cured, low w/b concrete Use of high alumina cement and calcareous (acid-soluble) aggregates lessens attack

Typical governing equations

1. Alkali reaction (selected)
Formation of (solid) secondary gypsum:

$$Ca(OH)_2 \rightarrow \underset{\text{Solid}}{Ca(OH)_2} + 2Na^+ + SO_4^{2-} \rightarrow \underset{\text{Solid secondary gypsum}}{CaSO_4 \cdot 2H_2O} + \underset{\text{Solution}}{2Na^+ + OH^-}$$

2. Strong acids (selected)
Direct attack on portlandite

$$Ca(OH)_2 + 2HA \rightarrow Ca^{2+} + 2A^- + 2H_2O$$

(Continued)

Table 4.1 (Continued) Summary of mechanical, physical, chemical, and biodeterioration mechanisms for concrete

Deterioration mechanism	Brief description and primary effect of deterioration	Nature of aggressive medium	Typical examples; likely occurrence and frequency	Avoidance or special precautions
3. Brackish water (selected) Magnesium sulfate attack on portlandite	$Ca(OH)_2 + MgSO_4 + 2H_2O \rightarrow \underset{Gypsum}{CaSO_4 \cdot 2H_2O} + \underset{Brucite}{Mg(OH)_2}$ $C-S-H + MgSO_4 + xH_2O \rightarrow CaSO_4 \cdot 2H_2O + Mg(OH)_2 + ySiO_2 \cdot H_2O$			
4. Sulfate attack (selected) Formation of gypsum (limited volume expansion)	$\underset{Gypsum}{Ca(OH)_2 + SO_4^{2-} \rightarrow CaSO_4} + H_2O$			
Calcium sulfate attack	$3CaSO_4 \cdot 2H_2O + 3CaO \cdot \underset{Tricalcium\ aluminate}{Al_2O_3} + 26H_2O \rightarrow \underset{Ettringite}{3CaO \cdot Al_2O_3 \cdot 3CaSO_4 \cdot 32H_2O}$			
Magnesium sulfate attack	$MgSO_4 + Ca(OH)_2 \rightarrow CaSO_4 + \underset{Brucite\ (insoluble)}{Mg(OH)_2}$			
5. Carbonic acid attack Formation of carbonates (lowers pH of concrete matrix)	$H_2CO_3 + Ca(OH)_2 \rightarrow CaCO_3 + 2H_2O$			
6. Biogenic H_2SO_4 attack Formation of gypsum (limited volume expansion)	$Ca(OH)_2 + H_2SO_4 \rightarrow CaSO_4 + 2H_2O$			

Details in this table are necessarily brief and not exhaustive. The interested reader is referred to additional references in "References" and "Further Reading."

Table 4.1 covers mechanisms, brief description and primary effect of deterioration, nature of aggressive medium, typical examples, likely occurrence and frequency, and avoidance or special precautions.

4.3.2 Mechanical and physical deterioration mechanisms

These mechanisms include abrasion, erosion, cavitation, freeze–thaw, salt crystallization, and cracking due to loading or thermal/hygral effects (e.g., temperature or moisture cycling). Cracking features in many deterioration mechanisms, either as a primary contributor or as a consequence. The resistance of concrete to mechanical deterioration mechanisms relates to its strength, toughness, and internal integrity, that is, absence of precracking or overloading, bond between paste and aggregate, nature and toughness of the aggregate itself (e.g., for abrasion), and the matrix microstructure, which must be dense and of low porosity. The presence of water during mechanical deterioration usually exacerbates the situation by softening the matrix and facilitating removal of the products of deterioration, such as in erosion. In general, the chemistry of the binder system is of less importance for mechanical or physical attack. Essential precautions are strong, dense concrete that is properly proportioned, free of macrocracks and defects such as excess voids, and containing aggregates that will enhance mechanical resistance if needed.

4.3.3 Chemical deterioration mechanisms

Many environments are chemically deleterious to concrete. These do not necessarily have to be highly concentrated chemical solutions. For example, pure water is very aggressive to concrete and yet is largely ion-free. The ion-hungry nature of such waters renders them aggressive, since concrete contains soluble species such as portlandite $(Ca(OH)_2)$ that can be leached from the matrix.

In most cases, chemical deterioration of concrete is characterized by interactions between the ions of the aggressive medium and the products of cement hydration, which also contain ions. These ion–ion interactions govern the nature and rate of deterioration as well as the resulting condition of the concrete. Figure 4.12 illustrates this and also shows typical examples of deterioration, while the following sections discuss various ionic interactions in deterioration.

4.3.3.1 Exchange reactions between external aggressive medium and components of hardened cement paste (ion exchange or substitution)

The products of cement hydration are stable in a high alkaline environment such as normal concrete, but this alkaline nature also renders it liable to

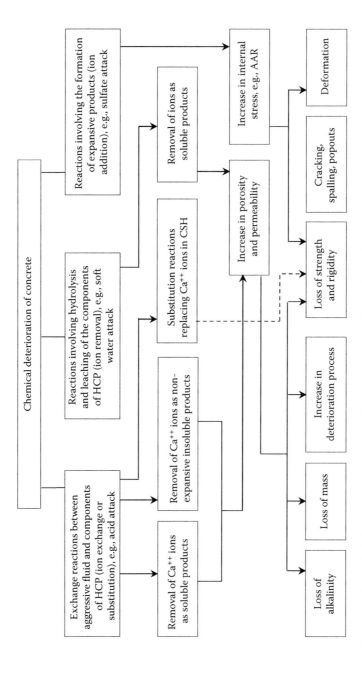

Figure 4.12 Forms of chemical deterioration and their effects in concrete. (Reproduced from Owens, G. (ed.), 2009, *Fulton's Concrete Technology*, Cement and Concrete Institute, Midrand, Figure 9.6, with permission from The Concrete Institute.)

deterioration by permitting the exchange or substitution of ions in the concrete with aggressive ions from the environment. A good example is attack by acids resulting in destruction of the cement paste matrix, where the acid anion is exchanged with the anions in the cement phases.

4.3.3.2 Hydrolysis and dissolution of the products of cement hydration (ion removal)

The products of cement hydration may in certain cases be dissolved and leached from the concrete matrix, causing the destruction of CSH gel with a consequential loss of concrete strength. The most readily soluble product is calcium hydroxide and so this is often the first substance to be dissolved or removed. Attack by soft or pure water is typical of this form of deterioration.

4.3.3.3 Expansive stresses caused by conversion of the products of hydration by agents (external or internal) (ion addition)

Normal concrete is porous and penetrable by external aggressive agents. Consequently, it is possible for such agents to enter the pore structure and react with the products of hydration. Certain aggressive agents have the ability to complex with the hydration products, forming new compounds internally in the concrete microstructure. Frequently, this results in internal expansive stresses, which cause cracking and destruction of the concrete matrix. Sulfate attack involving ion-addition (and ion-exchange) reactions with the hydroxide and aluminate phases, resulting in the formation of expansive products such as ettringite and gypsum, is a good example. Two special cases are as follows:

1. *Delayed ettringite formation*: This is an internal reaction caused by re-formation of ettringite at normal temperature, following high-temperature (>65°C) early curing (ettringite unstable above 65°C). This typically occurs in precast elements or large concrete masses where the early-age internal curing temperatures go too high, thus suppressing the formation of ettringite, which later re-forms in an expansive and destructive fashion.

2. *Alkali–aggregate reaction, or more commonly alkali–silica reaction*: The aggressive agents participating in the reaction are internal to the concrete, being a high alkali source (usually from the cement) and a reactive siliceous aggregate. This has the potential to form an alkali–silica gel, which in the presence of water particularly if from the exterior, can swell causing internal and external cracking and breakdown of the matrix—see Figure 4.10. (This is a complex topic; consult the references given in "Further Reading.")

Figure 4.13 Efflorescence on downstream face of a dam at Ocean Falls, British Columbia; the dam was constructed in 1925. (Courtesy of S. Mindess.)

Further aspects of Table 4.1 are briefly elaborated below.

- *Leaching*: This may lead to surface efflorescence from the deposition of salts leached out of the concrete, which then crystallize on the concrete surface on subsequent evaporation of the water, or by reaction with the CO_2 in the atmosphere. It is often referred to as lime "bloom," and can be seen frequently in situations where water can percolate through the concrete such as on the downstream faces of concrete dams (Figure 4.13) or when an exposed surface is alternately wetted and dried. It most commonly appears as a white deposit consisting primarily of $CaCO_3$, as the water dissolves some of the calcium hydroxide contained within the hydrated cement paste. Similarly, some of the calcium sulfate (gypsum) added to the cement to control setting may also dissolve in water and be drawn to the surface in the same way, where it too may leave a white deposit. Various other salts, whether internal or external to the concrete, have also been identified in deposits of efflorescence. Water temperature is also important in leaching, since calcium hydroxide is more soluble in cold than in warm water.

  Efflorescence is primarily an esthetic rather than a structural problem, but may indicate that a considerable amount of leaching is occurring within the concrete; in this case, there may be an increase in porosity and permeability and some strength loss, which may leave the concrete more open to other forms of chemical attack. In cases of very extensive leaching, the strength loss can be considerable. In an exhaustive review of the literature, Ekström (2001) reported strength losses of up to 50%, particularly where the water could seep through the concrete over a long period of time.

- *Freeze–thaw damage*: In addition to conventional freeze–thaw damage, cryogenic concrete, that is, concrete exposed to extremely low temperatures such as in tanks used for the storage of liquefied propane gas, can suffer similar damage. If the concrete has been well hydrated and cooled slowly, it will perform well with strengths two to three times higher than at room temperature, because of the freezing of the capillary moisture, and also a significantly increased elastic modulus. However, the concrete will undergo large, irreversible losses in strength after only one or two cycles of freezing to temperatures below about −70°C, and then thawing at room temperature.

- *Salt crystallization*: This problem is most commonly observed in the intertidal region in marine exposures where the concrete is subjected to regular cycles of wetting and drying. It may also occur if the concrete is in contact with groundwaters rich in salts. In this case, the groundwater may rise several hundred millimeters in the concrete through capillary action. If this rise extends above the ground, evaporation will concentrate just above ground level, and this is where the most severe damage will occur. This has been known to occur in some areas of the Middle East, and in other tropical regions. Damage has also been noted in floor slabs in southern California, in areas where the soil is rich in sulfates. Figure 4.14 is an example of such damage in Cuba, just a year or so after the concrete columns were constructed near the sea. This form of sulfate attack, due to the crystallization of sulfate salts, was in addition to the more "classical" form of sulfate attack.

Figure 4.14 Damage due to salt crystallization of concrete columns after about 1 year. (Courtesy of S. Mindess.)

The columns were constructed on the shore just meters from the sea near Havana, Cuba. The most effective way to limit such damage is by using a low w/c concrete. Other methods, such as building barriers into the structure to prevent capillary effects, can also be used.

4.3.4 Other factors to consider in chemical deterioration

The categories of chemical deterioration mechanisms above are useful in classifying and understanding such deterioration. However, actual situations are invariably more complex than given in simplified classification systems, and the following factors should be noted.

- Several deterioration mechanisms may operate simultaneously. For example, soft waters may contain dissolved carbon dioxide, which accelerates deterioration by leaching and acid attack.
- The moisture and temperature states of the concrete and environment should be considered. Chemical reactions are temperature dependent, but often the moisture state of the concrete is critical in deterioration rates. For example, carbonation takes place most rapidly in partially saturated concrete (at RH values of between about 50% and 70%), while chloride ingress requires saturated concrete for diffusion to occur.
- The microclimate of exposure can have a marked effect on the rate of deterioration. Whether the structure is exposed to rain and sun, or rain but no direct sun, or sheltered from rain and sun will greatly influence many deterioration processes. Likewise, two different aspects of the same structure or concrete element can have very different exposure microclimates, for example, sea-facing versus non-sea-facing aspects of the same building.
- Synergistic interactions of two or more deterioration mechanisms may occur. For example, chlorides penetrate more freely in carbonated concrete due to the reduced chloride-binding capacity of the cement.
- As mentioned previously, water—or moisture—is present in almost all chemical deterioration mechanisms. The presence of free water during deterioration of concrete very frequently exacerbates the mechanism, although not in all cases. Therefore, careful consideration must be given to the effects of water or the moisture state on deterioration. For example, saturated reinforced concrete permanently submerged in the ocean tends to be fairly well protected, while the same concrete in the splash/spray zone may deteriorate rapidly.

4.3.5 Steel corrosion in reinforced concrete

This is a very important mechanism of deterioration in reinforced concrete structures, and consequently is covered in some detail below.

4.3.5.1 Mechanism of steel corrosion in concrete

Corrosion of steel in concrete is an electrochemical process in which cathodic and anodic areas form on the steel surface due to electro-galvanic potential differences, with the concrete pore solution acting as an electrolyte. This results in the flow of local electric currents that constitute steel corrosion. The process can be described by the following equations (Bockris et al., 1981):

Anodic (oxidation) reaction:

$$Fe \rightarrow Fe^{2+} + 2e^- \tag{4.1}$$

Electrons released at the anode are consumed at the cathode by reduction. The commonly accepted reaction (based on the availability of O_2 and the pH at the steel surface) is the following:

$$2H_2O + O_2 + 4e^- \rightarrow 4OH^- \tag{4.2}$$

The electrical currents consist of electrons flowing through a conductor (the steel bar) between the anode and cathode, and ionic current flowing in the electrolytic solution—the pore solution in the concrete. The Fe^{2+} and $(OH)^-$ ions interact in the electrolyte solution, and in the presence of additional oxygen, they can form a range of iron oxides or "rust." The corrosion itself takes place at the anode with loss of steel surface as Fe^{2+} ions enter the electrolytic solution.

Figure 4.15 illustrates the corrosion process. Corrosion of steel in concrete is usually in the form of either microcell or macrocell corrosion. Macrocell corrosion is characterized by a clearly separated small localized

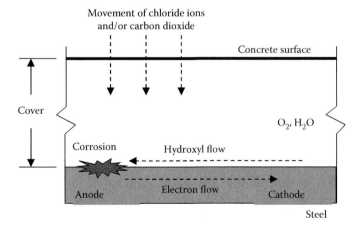

Figure 4.15 Schematic illustration of the corrosion process in concrete.

Figure 4.16 Pitting corrosion. (Courtesy of H. Beushausen.)

anode and a large cathode, typically occurring in chloride-induced corrosion or in concrete with low resistivity, with high local corrosion rates (pitting corrosion) and large steel cross-section reduction, as shown in Figure 4.16.

In microcell corrosion, the anodic and cathodic sites are closely adjacent, common in carbonation-induced corrosion or in concrete with high resistivity, for example, carbonated concrete.

4.3.5.2 Steel passivation and depassivation in concrete

Steel embedded in concrete is naturally protected by the high alkalinity of the cement matrix (pH > 12.5) and by the barrier effect of the concrete cover, which limits the oxygen, moisture, carbon dioxide, and chlorides required for active corrosion. This alkaline environment passivates the steel surfaces by the formation of a very thin (of the order of 10 nm) passive ferric oxide film (maghemite, γ-Fe_2O_3), resulting in negligible corrosion rates.

4.3.5.3 Corrosion products

The primary reaction of iron corrosion is the conversion of Fe to Fe^{2+} in the concrete. Numerous hydroxides, oxides, and oxide-hydroxides of the reacted iron form under different conditions of temperature, pressure, potential, and pH. The variations in volume change associated with individual iron oxides are illustrated in Figure 4.17. The formation of corrosion products causes expansive stresses in the concrete, with resulting cracking and spalling of the surrounding concrete.

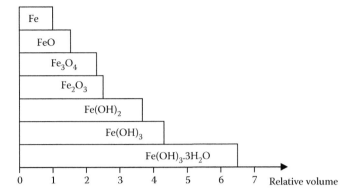

Figure 4.17 Relative volume of iron corrosion products. (Adapted from Liu, Y., 1996, Modelling the time-to-corrosion cracking of the cover concrete in chloride contaminated reinforced concrete structures, PhD dissertation, Virginia Polytechnic Institute and State University, Blacksburg, VA.)

4.3.5.4 Corrosion initiation

The passive layer can be disrupted or depassivated by ingress of corrosive species such as carbon dioxide and chlorides into concrete, leading to corrosion initiation. After depassivation, corrosion may occur and propagate depending on the availability of corrosion agents (specifically moisture and oxygen). Corrosion can also be initiated by stray currents and bacterial attack. However, the most common causes are carbonation- and chloride-induced corrosion, and these are briefly discussed.

Carbonation is a neutralizing reaction in the cement paste that reduces the pH from above 13 to less than 9, caused by the ingress of CO_2, which reacts with the calcium hydroxide in the concrete to form calcium carbonate in a two-step process given in Equations 4.3 and 4.4:

Formation of carbonic acid in the concrete pores:

$$H_2O + CO_2 \rightarrow H_2CO_3 \tag{4.3}$$

Reaction of carbonic acid with $Ca(OH)_2$ in the concrete:

$$H_2CO_3 + Ca(OH)_2 \rightarrow CaCO_3 + 2H_2O \tag{4.4}$$

This process depassivates the steel in contact with the carbonated zones. While the carbonation reaction is virtually self-sustaining because of the release of water, it tends to be limited by the increasing difficulty for the carbon dioxide to penetrate into the concrete.

In chloride-initiated corrosion, chloride ions that occur in the concrete either because they were admixed or after exposure to a chloride

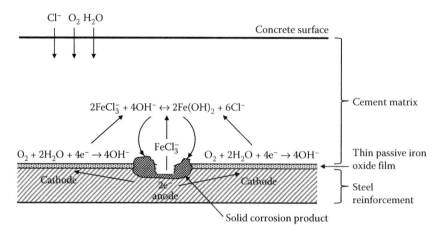

Figure 4.18 Corrosion of reinforcement in concrete exposed to chloride ions. (Adapted from Broomfield, J. P., 2007, *Corrosion of Steel in Concrete—Understanding, Investigation and Repair*, 2nd edition, Taylor & Francis, Oxford, UK.)

environment may depassivate the steel by locally breaking down the protective layer provided they are present in sufficient quantity. They are not consumed in the process but break down the passive layer and allow the corrosion process to proceed (Figure 4.18).

The process illustrated in Figure 4.18 results in very localized areas of the steel surface becoming anodic, with the passive surface forming the cathode. The resulting corrosion is in the form of localized pitting—see Figure 4.16. This kind of corrosion usually gives much higher corrosion currents than carbonation-induced corrosion and is particularly severe and pernicious.

As mentioned, the chloride ions must be present in sufficient concentration for depassivation to occur. This is referred to as the chloride threshold level (or critical chloride concentration) and is an input parameter in service life design and service life prediction models. It is not a single value valid for all types of concretes, steels, and environments, but is affected by different factors such as cover thickness, temperature, relative humidity, electrical potential of the reinforcement, chemistry of the binder, proportion of total chlorides to free chlorides, chloride to hydroxyl ion ratio, and condition of the steel surface. For practical purposes, it is often taken as 0.4% by mass of total binder,* with chlorides measured as the total (acid-soluble) chlorides in the concrete (Daigle et al., 2004). This threshold can be enhanced to provide additional stability to the film by the presence of

---

* Note that a variety of values for the threshold level are given in the literature, such as 0.1%, 0.2%, and 0.4% by mass of binder, or even higher. Each case needs to be considered on its own merits.

ions discharged from corrosion-inhibiting admixtures such as calcium nitrites—see Table 5.15 in Chapter 5.

4.3.5.5 Chloride binding in concrete

Not all of the chlorides in concrete are free or mobile and thus available for corrosion. Some are bound to the cement matrix, while free chlorides are dissolved in the concrete pore solution, their concentration reducing with time due to chloride binding, which removes chlorides from the pore solution through interaction with the cement matrix. Chlorides can be bound chemically by reacting with the aluminate phases (C_3A and C_4AF) to form calcium chloroaluminates, or they may be physically bound to hydrate surfaces by adsorption. All cements bind chlorides to some degree and this strongly influences the rate at which chlorides penetrate into the concrete from an external source.

Bound chlorides are in chemical equilibrium with the free chlorides. They can be released under certain conditions to become free chlorides again, such as when chloride-contaminated concrete subsequently carbonates causing a release of the bound chlorides. Thus, bound chlorides also present a corrosion risk. The nature of the cement chemistry is important in chloride binding, and the choice of binder type is critical. In this respect, blended binders containing fly ash or slag are superior due to their elevated chloride-binding properties.

4.3.5.6 Corrosion rate in concrete

The overall rate of the electrochemical process is determined by the slowest of the processes involved: anodic chemical reaction, cathodic chemical reaction, and ionic current flow. The cathodic reaction rate depends upon the availability of oxygen diffusing through the concrete cover, while the ionic current rate depends upon the electrical resistance of the concrete cover (Figure 4.19).

The overall corrosion process of steel in concrete is usually described in terms of a corrosion versus time diagram, showing the depassivation stage and the propagation stage, which occurs afterwards (Figure 4.20). The rates of corrosion in the depassivation stage are usually controlled by the availability of oxygen and the electrical resistance of the concrete. Once sufficient rust has accumulated around the steel, cracking of the concrete cover takes place due to the expansive forces exerted by the voluminous rust, and the corrosion rates are further accelerated.

4.3.5.7 Corrosion of steel in cracked concrete
         (Otieno et al., 2015a,b)

In cracked concrete, corrosion starts either in the crack zone or in zones immediately adjacent to the crack. There are two different corrosion

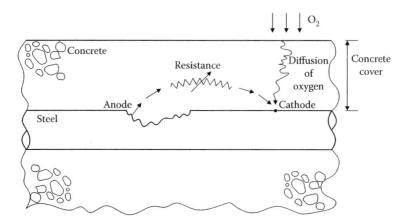

Figure 4.19 Rate determining factors of the electrochemical corrosion process of steel in concrete. (Reproduced from Bentur, A., Diamond, S., and Berke, N. 1997, *Steel Corrosion in Concrete, Fundamentals and Civil Engineering Practice,* E&FN SPON/ Chapman and Hall, London, UK, with permission from Taylor & Francis.)

mechanisms that can exist theoretically in the region of cracks (Schießl and Raupach, 1997):

- *Mechanism I*: Both the anodic and cathodic sites are very small and located closely to each other (microcell corrosion) in the crack zone.
- *Mechanism II*: The anode is the reinforcement in the crack zone, while the passive steel surface between the cracks forms the cathode.

Corrosion in cracked concrete is governed by further complicating factors compared to uncracked concrete, mainly the nature and extent of cracking. Cracking greatly increases the penetrability of the cover layer, substantially reduces the period of time before active corrosion occurs, and thereafter tends to accelerate the rate of corrosion, thereby reducing the service life of the structure. The use of SCMs such as fly ash, condensed silica fume, or ground-granulated blast-furnace slag, can reduce the effect of cracks to some extent. Nevertheless, cracked reinforced concrete in corrosion-prone zones such as marine areas or chloride-contaminated environments presents a particular challenge in terms of creating a durable structure.

To limit the corrosion rate of reinforcing steel in structures located in corrosion-prone zones, certain inter-linking parameters that include concrete quality, cover depth, binder type, and w/b ratio need to be considered during the design stage. A comparison of the influence of some of these parameters on the corrosion rate of reinforcing steel in uncracked and cracked (0.4 mm crack) concrete is illustrated in Figure 4.21. This covers a range of concretes based on plain Portland cement (PC), and blends with SCMs such as fly ash (FA) and slag (SL), with w/c being either 40% or 55%.

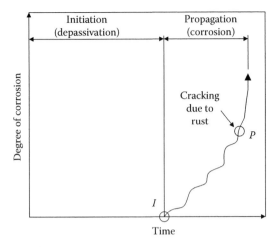

Figure 4.20 Modeling of the corrosion process of steel in concrete. (Reproduced from Bentur, A., Diamond, S., and Berke, N., 1997, *Steel Corrosion in Concrete, Fundamentals and Civil Engineering Practice,* E&FN SPON/Chapman and Hall, London, UK, with permission from Taylor & Francis.)

Figure 4.21 confirms that the corrosion rate of embedded steel in reinforced concrete increases with crack width and lower cover, and that the use of SCMs can decrease the corrosion rate.

4.3.5.8 Electrochemical corrosion and nondestructive testing techniques

The driving forces for creating the electrochemical process of corrosion can be presented in electrical terms, which form the basis for various nondestructive testing techniques. This aspect is covered in Chapter 8.

4.4 DESIGN AND CONSTRUCTION CONSIDERATIONS TO MINIMIZE CONCRETE DETERIORATION

There are accepted technologies to minimize deterioration in concrete structures. Newer approaches are becoming available to assist designers and constructors in achieving durable structures. This section briefly outlines established techniques to help ensure resistance against deterioration of concrete.

4.4.1 Concrete mix design and materials selection

First, attention must be paid to selection of the concrete mix constituents and mix proportions, in order to minimize concrete penetrability and

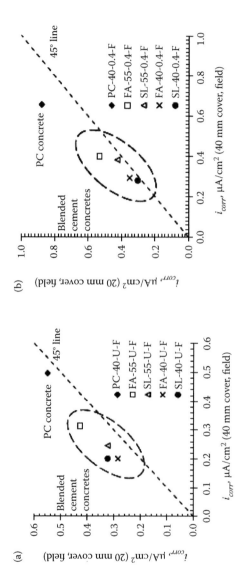

Figure 4.21 Influence of binder type and cover depth on the corrosion rate of reinforcing steel in concrete exposed in a marine environment. (a) Uncracked concrete, (b) 0.4 mm cracked concrete. (Reprinted from Cement and Concrete Research, 79(2016), Otieno, M., Beushausen, H., and Alexander, M. G., Chloride-induced corrosion of steel in cracked concrete—Part I: Experimental studies under accelerated and natural marine environments, 373–385, Copyright 2015, with permission from Elsevier.)

maximize chemical resistance. This requires lowering the water content of the concrete as far as practically possible, as well as careful aggregate and admixture selection. A suitable w/b ratio must be chosen, bearing in mind that modern cements tend to produce higher strengths than earlier ones at the same w/b ratios, which produces concrete of adequate strength but higher porosity and permeability.

The chemical resistance of the concrete derives mainly from the binder system, which must be selected to "match" the environment. For example, carbonating conditions require a cement with a high CaO content, while chloride resistance requires cements rich in aluminate phases to "bind" the chloride ions. Also, research is increasingly showing that the w/b ratio is more important than the binder content per se (Wassermann et al., 2009).

4.4.2 Reinforcement cover

Inadequate cover of reinforcing bars is probably the single most important reason for premature deterioration of reinforced concrete, arising either by design and detailing faults or through negligence during construction. Simply observing the expedient of adequate cover would help to resolve the great majority of corrosion problems. Also, it is not only essential to specify correct cover, but even more importantly to actually achieve the specified cover during construction. Codes of practice for concrete design and construction give guidance on steel cover required for different aggressive environments, although these may be overly conservative, or even occasionally nonconservative with regard to actual performance. For this reason, there are international trends to move away from the prescriptive rules of the past and move toward performance-based design and construction philosophies that include the dual effects of concrete cover depth and quality of cover concrete, as discussed in Chapters 5 and 6 of this book.

Consistently achieving the required cover depths on construction sites is not a simple matter. Attention must be paid to providing good quality, dense cover blocks, placed at a spacing close enough to ensure that the reinforcing cage does not move unduly during concreting, allowing for tolerances on element dimensions and reinforcing steel bending dimensions (see also Chapter 10).

4.4.3 Curing of concrete

Curing is intended to ensure that the hydration reactions progress to as great a degree as possible, ensuring a dense microstructure with minimal porosity and maximum pore-discontinuity. This requires attention to the moisture and temperature environment in which the fresh concrete hardens and develops its properties, particularly in the critical first hours and days of its existence. There are many useful guides to good curing available

(CSA A 23.1; ACI 308R; BS EN 13670); the challenge is actually achieving good curing during construction. Modern construction practices such as fast-track construction often result in curing being neglected. Construction methods such as slip-forming and bridge-jacking require special attention to curing because of the short time in which the concrete remains in the formwork. Membrane-forming curing compounds can be convenient and effective means of curing concrete, but subsequent evaluation of the quality of the cover has indicated that required properties have not always been achieved.

As with specifying concrete cover, there is a move away from prescriptive means of achieving curing to performance-based approaches such as measuring the quality of the cover concrete after curing to ensure adequate cover properties. This is discussed in Chapter 6; other references are Beushausen and Fernandez (2015) (RILEM TC 230—PSC Report) and Owens (2009) (Chapter 9).

4.5 CONCLUDING COMMENTS

Concrete deterioration is a serious problem for the durability of many concrete structures. Various deterioration mechanisms affect concrete structures, and these have been reviewed in this chapter. Steel corrosion in reinforced concrete probably remains the most serious deterioration mechanism currently afflicting concrete structures, and this needs special attention in design, construction, and operation. Chapters 5 and 6 in this book deal with newer approaches to durability design and specification, such as performance-based approaches and durability indicators. These newer approaches show promise in helping to improve the longevity of concrete structures.

REFERENCES

ACI 308R-01, 2008, *Guide to Curing Concrete*, American Concrete Institute, Farmington Hills, MI.

Aldea, C. M., Shah, S. P., and Karr, A., 1999, Permeability of cracked concrete, *Materials & Structures*, 32, 370–376.

Alexander, M. G., Bertron, A., and De Belie, N. (eds.), 2013, Performance of cement-based materials in aggressive aqueous environments, State-of-the-Art Report, RILEM TC 211–PAE, Springer, Dordrecht, 449pp.

Al Hashem, A., 2011, Corrosion in the Gulf Cooperation Council (GCC) states: Statistics and figures, Proceedings of the Corrosion UAE, Abu Dhabi, February 27–March 1.

Andrade, C., 1993, Calculation of chloride diffusion coefficients in concrete from ionic migration measurements, *Cement and Concrete Research*, 23(3), 724–742.

ASTM E632-82, 1996, *Standard Practice for Developing Accelerated Tests to Aid Prediction of the Service Life of Building Components and Materials*, American Society for Testing and Materials, Michigan. Standard withdrawn in 2005.

Bentur, A., Diamond, S., and Berke, N., 1997, *Steel Corrosion in Concrete, Fundamentals and Civil Engineering Practice*, E&FN SPON/Chapman and Hall, London, UK.

Beushausen, H. and Fernandez, L. (eds.), 2015, Performance-based specifications and control of concrete durability, State-of-the-Art Report, RILEM TC 230—PSC, Springer, Dordrecht.

Biczók, I., 1964, *Concrete Corrosion and Concrete Protection*, Hungarian Academy of Sciences, Budapest, Hungary.

Bockris, J., Conway, B., Yeager, E., and White, R. (eds.), 1981, *Comprehensive Treatise of Electrochemistry, Electrochemical Processing*, Volume 4, Springer, New York, NY.

Broomfield, J. P., 2007, *Corrosion of Steel in Concrete—Understanding, Investigation and Repair*, 2nd edition, Taylor & Francis, Oxford, UK.

BS EN 13670, 2009, *Execution of Concrete Structures*, British Standards Institution, London, UK.

BS 6349-1-4:2013, 2013, *Maritime Works. General. Code of Practice for Materials*, British Standards Institution, London, UK.

Claisse, P. A., 2005, Transport properties of concrete, *Concrete International*, 27(1), 43–48.

Concrete Society TR22, 2010, *Non-Structural Cracks in Concrete*, 4th edition, Concrete Society, Surrey, UK.

CSA A 23.1, 2014, *Concrete Materials and Methods of Concrete Construction/ Test Methods and Standard Practices for Concrete*, Canadian Standards Association, Toronto, Canada.

Daigle, L., Lounis, Z., and Cusson, D., 2004, Numerical prediction of early-age cracking and corrosion in high performance concrete bridges—Case study, *Proceedings of the Annual Conference of the Transportation Association of Canada Innovations in Bridge Engineering*, Québec, September 19–23, pp. 1–20.

Ekström, T., 2001, *Leaching of concrete*, Report TVBM-3090, Lund University, Lund, Sweden.

Gjørv, O. E., 2009, *Durability Design of Concrete Structures in Severe Environments*, Taylor & Francis, Oxford, UK.

Heckroodt, R. O., 2002, *Guide to the Deterioration and Failure of Building Materials*, Thomas Telford, London, UK.

Liu, Y., 1996, Modelling the time-to-corrosion cracking of the cover concrete in chloride contaminated reinforced concrete structures, *PhD dissertation*, Virginia Polytechnic Institute and State University, Blacksburg, VA.

Nganga, G., Alexander, M. G., and Beushausen, H., 2013. Practical implementation of the durability index performance-based design approach, *Construction and Building Materials*, 45, 251–261.

Otieno, M., 2014, The development of empirical chloride-induced corrosion rate prediction models for cracked and uncracked steel reinforced concrete structures in the marine tidal zone, PhD dissertation, University of Cape Town, Cape Town, South Africa.

Otieno, M., Beushausen, H., and Alexander, M. G., 2015a, Chloride-induced corrosion of steel in cracked concrete—Part I: Experimental studies under accelerated and natural marine environments, *Cement and Concrete Research*, 79(2016), 373–385.

Otieno, M., Beushausen, H., and Alexander, M. G., 2015b, Chloride-induced corrosion of steel in cracked concrete—Part II: Corrosion rate prediction models, *Cement and Concrete Research*, 79(2016), 386–394.

Owens, G. (ed.), 2009, *Fulton's Concrete Technology*, 9th edition, Cement and Concrete Institute, Midrand, South Africa.

Schießl, P. and Raupach, M., 1997, Laboratory studies and calculations on the influence of crack width on chloride-induced corrosion of steel in concrete, *ACI Materials Journal*, 94(1), 56–62.

Soutsos, M. (ed.), 2010, *Concrete Durability*, Thomas Telford, London, UK.

Wassermann, A., Katz, A., and Bentur, A., 2009, Minimum cement content requirements: A must or a myth? *Materials and Structures*, 42, 973–982.

FURTHER READING

ACI 201.2R, 2008, *Guide to Durable Concrete*, American Concrete Institute, Farmington Hills, MI.

ACI 224R-01, 2001, *Control of Cracking in Concrete Structures*, American Concrete Institute, Farmington Hills, MI.

Arya, C. and Ofori-Darko, F. K., 1996, Influence of crack frequency on reinforcement corrosion in concrete, *Cement and Concrete Research*, 26(3), 333–353.

Basson, J. J., 1989, *Deterioration of Concrete in Aggressive Waters: Measuring Aggressiveness and Taking Counter Measures*, Portland Cement Institute, Midrand, South Africa.

Birt, J. C., 1981, *Curing Concrete—An Appraisal of Attitudes, Practices and Knowledge*, 2nd edition, CIRIA, London, UK.

Böhni, H. (ed.), 2005, *Corrosion in Reinforced Concrete Structures*, Woodhead Publishing Ltd, Cambridge, UK.

BRE, 2005, *Concrete in Aggressive Ground*, Parts 1–4, 3rd edition, Building Research Establishment, Bracknell, UK.

Buenfeld, N. R. and Yang, R., 2001, Onsite curing of concrete: Microstructure and durability, CIRIA Report C530, CIRIA, London, UK.

Concrete Society TR31, 2008, *Permeability Testing of Site Concrete: A Review of Methods and Experience*, Concrete Society, Surrey, UK.

Concrete Society TR44, 1995, *The Relevance of Cracking in Concrete to Corrosion of Reinforcement*, Concrete Society, Surrey, UK.

Czernin, W., 1980, *Cement Chemistry and Physics for Civil Engineers*, 2nd edition, Bauverlag, Berlin, Germany.

Detwiler, R. J. and Taylor, P. C., 2005, *Specifiers Guide to Durable Concrete: Engineering Bulletin 221*, Portland Cement Association, Skokie, IL.

Eglinton, M. S., 1987, *Concrete and Its Chemical Behaviour*, Thomas Telford, London, UK.

Harrison, T. A., 1981, Early-age thermal crack control in concrete, CIRIA Report 91, CIRIA, London, UK.

International Federation for Structural Concrete, *fib*, 2010, *The fib Model Code for Concrete Structures*, Ernst & Sohn, Berlin, Germany.

Mehta, P. K. and Monteiro, P., 2006, *Concrete—Microstructure, Properties, and Materials*, McGraw-Hill, New York, NY.

Otieno, M. B., Alexander, M. G., and Beushausen, H., 2010a, Corrosion in cracked and uncracked concrete—Influence of crack width, concrete quality and crack re-opening, *Magazine of Concrete Research*, 62(6), 393–404.

Otieno, M. B., Alexander, M. G., and Beushausen, H., 2010b, Suitability of various measurement techniques for assessing corrosion in cracked concrete, *ACI Materials Journal*, 107(5), 481–489.

Portland Cement Association, 1997, *Effects of Substances on Concrete and Guides to Protective Treatments: Concrete Information Pamphlet*, Portland Cement Association, Skokie, IL.

Rajabipour, F., Giannini, E., Dunant, C., Ideker, J., and Thomas, M., 2015, Alkali–silica reaction: Current understanding of the reaction mechanisms and the knowledge gaps, *Cement and Concrete Research*, 76, 130–146.

Raupach, M., 1996, Chloride-induced macrocell corrosion of steel in concrete—Theoretical background and practical consequences, *Construction and Building Materials*, 10(5), 329–338.

Scott, A. N. and Alexander, M. G., 2007, The influence of binder type, cracking and cover on corrosion rates of steel in chloride-contaminated concrete, *Magazine of Concrete Research*, 59(7), 495–505.

Suzuki, K., Ohno, Y., Praparntanatorn, S., and Tamura, H., 1990, Mechanism of steel corrosion in cracked concrete, *Proceedings of the Third International Symposium on Corrosion of Reinforcement in Concrete Construction*, Warwickshire, UK, May 21–24, pp. 19–28.

Thomas, M., 2011, The effect of supplementary cementing materials on alkali-silica reaction: A review, *Cement and Concrete Research*, 41(12), 1224–1231.

Chapter 5

Durability specifications, limit states, and modeling

5.1 BASIC CONCEPTS

Traditionally, the procedures for designing concrete mixtures, and indeed for designing concrete structures themselves, have focused primarily on compressive strength and other mechanical properties; durability issues received only secondary consideration. However, more recently, the focus has shifted to designing first for durability, at least for major structures. This change has come about for several reasons:

- Using modern concrete technology, it is relatively easy to achieve high strengths; however, high strength does not necessarily translate to high durability.
- With the increasing emphasis on the sustainability of concrete structures, it has become apparent that increasing the durability (and hence the service life) of concrete structures is much more effective in reducing the carbon footprint than merely increasing the strength, or replacing some Portland cement with other pozzolanic materials.
- Problems with concrete structures are rarely due to deficiencies in strength; premature distress or failure of concrete structures is almost always initiated by inadequate durability. It should be noted, however, that "durable" concrete does not imply "maintenance-free" concrete. Inherent in the notion of durability is the obligation to inspect the concrete on a regular basis, and to carry out such (reasonable) maintenance and repairs as required to keep the structure functional as long as possible.

In order to ensure durable concrete for any particular application, it is essential that the concrete specifications be clear and unambiguous. There is little point in specifying a service life of, say, 100 years, if there is no way of specifying at the same time the predictive models, the concrete tests to be used, the schedule of inspections, and the circumstances under which repairs or rehabilitation is to be carried out. While this may seem to be quite straightforward in principle, the reality is much more complicated.

In essence, we are trying to predict long-term behavior on the basis of short-term tests, in conjunction with predictive models that inevitably are simplifications of the real situation to which the concrete will be subjected. This can lead to a myriad of problems.

Durability of a structure implies that it remains fit for use for its intended purpose in the environmental conditions to which it is exposed over its entire service life, including anticipated maintenance, but without the need for major repair. Design for durability has become an inherent part of standards and design codes. We are currently at a period in which the approach in the codes is gradually changing from a general prescriptive one to performance-based durability design and testing. In order to provide a coherent overview of the current state of affairs and the expected future developments, there is a need to clearly define the concepts of durability design and the means to achieve the expected long-term performance. These concepts and the derived definitions are embedded in standards and codes and are specified in a variety of documents, such as fib model code for service life design (fib, 2006), fib model code 2010, ISO 13823 (2008), ISO 16204 (2012), ACI 365.1R-00 (2000), and Eurocode 2: EN (1992).

It should be noted that there is some overlap between the concepts discussed in this chapter, and those in Chapter 6 of this book. Overall, this provides a richer understanding of these important topics for durability.

5.1.1 Performance versus prescriptive specifications

The concrete industry is currently in the process of moving, very gradually, from prescriptive to performance specifications in the design of concrete mixtures. At the moment, we still rely heavily upon *prescriptive* specifications, which generally include requirements such as maximum w/b ratios, minimum cementing materials contents, cement types, limitations on the types and amounts of both chemical and mineral admixtures, limitations on filler materials in the cement itself, and so on. They are also primarily strength based. Though modern concrete design codes do, of course, contain provisions for durability, the primary criterion for judging the adequacy of concrete for most applications remains its compressive strength (f_c'). Unfortunately, particularly in terms of durability, this is far from a guarantee of concrete adequacy.

However, the concrete industry has in recent years become much more sophisticated, and much more focused on the issue of sustainability. Unfortunately, one consequence of the prescriptive approach to mix design is that it inhibits the most efficient use of the many materials now available to make up a concrete mixture. It is therefore time to make much more extensive use of *performance* specifications. Such specifications, properly written, would permit concrete producers to be more imaginative and innovative in their choice of materials, such as less common supplementary cementing materials (SCMs), admixtures, blended cements, polymers, fibers, mineral fillers, and so on. They would also be a means for introducing durability concerns more explicitly into the design of concrete mixtures.

The intent of performance specifications is to prescribe the required properties of the concrete in both the fresh and hardened states, but without saying explicitly how they are to be achieved. A typical definition of such a specification is provided in the Canadian Standard CSA A23.1 (2004):

> A performance concrete specification is a method of specifying a construction product in which a final outcome is given in mandatory language, in a manner that the performance requirements can be measured by accepted industry standards and methods. The processes, materials or activities used by the contractors, subcontractors, manufacturers and materials suppliers are then left to their discretion. In some cases, requirements can be referenced to this Standard [CSA A23.1], or other commonly used standards and specifications, such as those covering cementing materials, admixtures, aggregates, or construction practices.*

The difference between prescriptive and performance specifications is stark. In the former, the owner (i.e., his or her architectural and engineering consultants) specifies the concrete mix proportions, the materials to be used, the properties of the fresh and hardened concrete, and any other requirements. The owner thus assumes full responsibility for the concrete; the materials supplier and the contractor simply have to follow these specifications. The use of performance specifications, on the other hand, requires the owner, the contractor, and the materials supplier to cooperate with each other from the outset. The owner specifies all of the required concrete properties, the supplier assumes responsibility for delivering the appropriate concrete to the jobsite, and the contractor assumes responsibility for placing and curing the concrete.

However, the move from prescriptive to performance specifications will not be quick or easy. There are several reasons for this:

- The conservatism of the concrete industry as a whole.
- Many concrete producers do not have the properly trained personnel to provide the necessary technical advice to specifiers and engineers, or to deal with complex quality control issues.

---

* Note: With the permission of the Canadian Standards Association (operating as CSA Group), material is reproduced from CSA Group Standards, *A23.1-14/A3.2-14—Concrete Materials and Methods of Concrete Construction/Test Methods and Standard Practices for Concrete*, which is copyrighted by CSA Group, 178 Rexdale Blvd., Toronto, ON, M9W 1R3. This material is not the complete and official position of CSA Group on the referenced subject, which is represented solely by the standard in its entirety. While use of the material has been authorized, CSA Group is not responsible for the manner in which the data is presented, nor for any interpretation thereof. No further reproduction permitted. For more information or to purchase standards from CSA Group, please visit http://shop.csa.ca/ or call 1-800-463-6727.

- Perhaps of most importance is that there is currently a lack of quick, reliable durability tests. If we are to move away from using the 28-day compressive strength as the primary arbiter of concrete quality, such tests are essential. Until such tests are developed and verified in practice, we may still have to rely on the "equivalent performance" concept as described below.
- Finally, with the shared responsibility envisaged by the concept of performance specifications, there would have to be some way of assigning responsibility for the concrete quality, particularly in cases in which the job requirements are not met.

5.1.2 Predictive models of service life

The prediction of service life is of great importance to owners of major structures and civil engineering infrastructure (dams, bridges, sewer systems, and so on), so that they can develop timelines and budgets for future maintenance and eventual replacement. In the early days of reinforced concrete structures, it was assumed that if the right steps were taken in the design and production of the concrete, it would then perform satisfactorily for the life of the structure, however long that might be (Thompson, 1909):

> When well designed and properly constructed, a reinforced concrete structure will be safe for all time, since its strength increases with age, the concrete growing harder and the bond with the steel becoming stronger. ...a concrete structure as stated above grows stronger with time and its life is measured by ages rather than years.

This was written, of course, long before such issues as steel corrosion, freeze–thaw resistance, sulfate attack, alkali–aggregate reactions (AARs), and so on were even imagined, much less understood.

With the increasing use of high-performance concrete, and the realization that a longer service life would not only be more economical in the long run, but would also lead to a reduced carbon footprint, more and more owners are now specifying a service life of 100 years, or even longer. This is, however, still largely uncharted territory, as the tools to make such long-term predictions have not yet been fully developed. There are currently a number of models that purport to predict long-term behavior for certain environmental exposure conditions, some of which are listed in Table 5.1. The most basic models are intended to predict the time to corrosion of the reinforcing steel for structures exposed to chlorides, assuming that the fluid transport is by diffusion alone. More sophisticated models also deal with some (but not all) other types of chemical exposure and may consider additional fluid transport mechanisms. As yet, none of the existing models can deal with the effects of construction or design defects. Also, service life

Table 5.1 Models for prediction of service life

Model	Characteristics	Reference
LIFE-365®	Chloride diffusion model, based on Fick's law. Semi-probabilistic. Provides life-cycle cost analysis	Free software. www. Life-365.org
STADIUM®	Multi-ionic model, based on Ernst–Planck equation. Provides chloride ingress rate and corrosion initiation. Also provides carbonation and sulfate profiles. Full probabilistic	Proprietary software. www. simcotechnologies.ca
fib Bulletin 34	Based on Fick's second law. Deals primarily with chloride ingress and carbonation. Used in *fib* Model Code 2010. Full probabilistic	Open source bulletin. ISBN: 978-2-88394-074-1
ConcreteWorks	Based on Fick's law. Predicts strength, chloride ingress, thermal cracking	www.texasconcrete-works.com; Folliard et al. (2008)
ClinConc	Chloride diffusion model	Tang (2008)
LIFEPRED	Based on Fick's law	Andrade and Tavares (2012)

design criteria cannot yet adequately or fully consider the trade-off between initial cost and downstream maintenance and rehabilitation costs.

Of course, it is well known that fluid ingress into concrete does not occur due to diffusion alone. There are other transport mechanisms that must be considered, primarily, *permeability* due to a pressure head or thermal gradient; *sorption* due to repeated cycles of wetting and dying; and *wick action* if there is a drying surface. Above all, the presence of cracks will always exacerbate the ingress of fluids. In reality, several of these mechanisms may be acting at the same time, so that it can be very difficult to predict the ingress of harmful fluids in severe exposure conditions, or when the concrete is subjected to several destructive mechanisms simultaneously (e.g., sulfate attack + chloride corrosion + freeze–thaw cycles).

There are several other problems to the above "time to corrosion" models. The assumptions and the algorithms used in these models are difficult to verify independently, particularly since they often involve proprietary software and "in-house" databases. This makes any independent review of the validity of these models difficult. The models also fail to deal with many other types of deterioration, such as AAR, scaling, D-cracking, and so on. So, using this approach, we end up trying to design structures to satisfy a numerical value of service life as calculated from the various models, in order to meet subjectively defined levels of durability and service life. There is clearly a need for a much more robust approach to designing for service life based on durability considerations. It might be more helpful if future research and development focused less on developing precise numerical models, and more on defining exposure conditions and the related probability distributions. There should also be more focus on correlating laboratory tests with field performance and the effectiveness of mitigation measures.

5.1.3 Service life design codes and standards

At this time, there are no North American codes or standards for service life design. There are a number of reasons for this:

- The various jurisdictions and agencies in North America do not have a unified approach to tests and specifications, or to how to use the results so obtained.
- Similarly, there are different approaches to how exposure conditions should be defined, and how they are to be dealt with.
- As stated above, there are many different approaches to the prediction of service life, and indeed how to specify in-service performance in project specifications. Statements such as "...*shall perform its design function without significant repairs, rehabilitation or replacement*" or "*service life shall be defined as the time in service until spalling of concrete occurs*" are too open to interpretation.
- It is difficult for a designer or owner to make an independent verification of the analysis used in the various models; this leads to a lack of confidence in the output of these models.

In spite of these issues, however, there are available several international service life design codes, including *fib* Bulletin 34, *Model Code for Service Life Design* (2006), and ISO 16204 (2012), *Service Life Design of Concrete Structures* (2012). (These two are quite similar, since ISO 16204:2012 is based in large part on *fib* Bulletin 34.) The basic idea behind these codes is to establish an approach to concrete design to avoid deterioration of the concrete due to environmental effects, to stand alongside the usual structural design codes. (It should be noted, however, that these codes are not well adapted to North American materials and practices.) The design process involves four steps:

1. *Quantify the deterioration mechanism* using realistic models that can describe the process with sufficient accuracy either physically or chemically. This would require a combination of laboratory and field observations, and sufficient data to provide statistical parameters (mean values, scatter, etc.). For some deterioration mechanisms, generally accepted mathematical models exist, which take into account both the material and the environmental parameters (mix design, w/c ratio, temperature, relative humidity, etc.). Such mechanisms include carbonation-induced corrosion of the reinforcing steel, chloride-induced corrosion, and freeze–thaw attack. For other mechanisms, such as

AAR, sulfate attack, or other chemical attack, such models do not yet exist, and would need to be developed in order to fit neatly into this design procedure. As well, cases in which several deterioration mechanisms are acting simultaneously, or where there is already cracked or otherwise damaged concrete, cannot yet be dealt with.

2. *Define the limit states* appropriate to the particular project. These might include such factors as depassivation of the reinforcement due to carbonation, cracking due to corrosion of the steel, spalling of the concrete, collapse due to loss of cross section of the reinforcement, and so on.
3. *Calculate the probability* that the limit states defined in Step 2 will be exceeded, by applying the models adopted in Step 1.
4. *Define the type of limit state*, with reference to the limit states defined in Step 1. The options are a *serviceability limit state* (SLS), which refers to the point beyond which the specified service requirements are no longer met, and an *ultimate limit state* (ULS), which is associated with collapse or similar forms of structural failure.

In order to produce a rational service life design with a given level of reliability in accordance with the above process, ISO 16204:2012 provides two different strategies:

Strategy 1, which is based on designing the structure to *resist* deterioration, permits three different levels of sophistication, depending on the importance of the structure:

- The full probabilistic method is basically *performance* based; it requires the existence of probabilistic models for the strength (resistance) of the structure as designed, for the loads and exposure conditions, and for the geometry of the structure. It also requires a clear definition of the limit states beyond which the structure fails or cannot meet serviceability requirements.
- The partial factor design method is also largely *performance* based. It is a deterministic approach, in which the scatter in both strength (as predicted from the design calculations) and environmental loads is taken into account by partial safety factors. This too requires a definition of the limit states.
- The "deemed-to-satisfy" method is similar to the design methods currently in use. That is, it involves a mixture of theory, laboratory tests, and field experience, but is not based directly on proper physical and chemical models, nor is it probabilistic. It is largely a *prescriptive* approach, and is described in detail in Section 5.6 of this chapter.

Strategy 2 is based on designing the structure completely to *avoid* degradation processes. This could be based on protective technologies such as cladding and membranes which isolate the structure from the environment, or using components which are nondegradable, such as stainless steel to replace conventional steel reinforcement in concrete. This approach calls for specifications that are prescriptive in nature. It should be emphasized that for both the "deemed-to-satisfy" and "avoidance" approaches, there is a lack of reliable test methods that correlate with field experience. As well, neither the control nor the avoidance measures are correlated to any specific length of service life.

A flow chart outlining the different design schemes is presented in Figure 5.1.

5.1.4 Equivalent performance concept

There is increasing interest in concretes made with what may be termed *alternative* concrete compositions, due to the use of many different types of cements (not always based on Portland cement) and SCMs. These concretes often fall outside of the standardized (prescriptive) specifications for concrete, in terms of such parameters as w/b ratio, minimum cement contents, and so on. In order to evaluate such concretes, EN 206-1 (2013) provides a test methodology, the *Equivalent Concrete performance Concept*. This is a comparative test, in which the "alternative" concrete is compared to a reference concrete whose composition is generally accepted to be suitable for the application in question. What is key in this approach is the selection of the reference concrete. The reference concrete mixture is based on the minimum composition requirements of EN 206-1 for the environmental class appropriate to the application. There are, however, two additional criteria: the w/b ratio is reduced by 0.05 compared to the limiting value for the particular exposure class, and the binder content is increased by 5%. These criteria are intended to ensure that the reference concrete is considerably better than the minimum composition given by EN 206-1. Generally, both concretes are made with the same aggregate.

The specimens of the reference concrete and the "new" concrete are then compared with one another in terms of these properties:

- Compressive strength
- Resistance to carbonation
- Chloride penetration
- Freeze–thaw resistance (in the presence of deicing salts)
- Other possible requirements, depending upon the particular application, which might include such things as scaling resistance, resistance against seawater, sulfate resistance, acid attack, and ASR

The alternative concrete must be as good as or better than the reference concrete in all of these tests. This approach has been discussed in more detail by de Schutter (2009) and Linger et al. (2014).

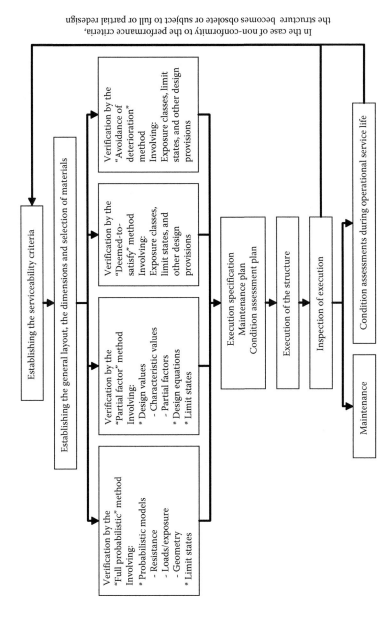

Figure 5.1 Flow chart of the different approaches for service life design. (Adapted from ISO 16204, 2012, *Durability—Service Life Design of Concrete Structures*; Reproduced with the permission of the South African Bureau of Standards [SABS].) The South African Bureau of Standards is the owner of the copyright therein and unauthorized use thereof is prohibited. The standard may be obtained at the following: www. store.sabs.za

5.1.5 Durability tests

Over the years, a great many durability tests have been developed for concrete (e.g., Gehlen and Rahimi, 2011); a number of these are described in detail in Chapters 7 and 8. In general, these tests can be categorized as the following:

1. *Freezing and thawing resistance:* These are among the oldest durability tests and involve some prescribed cycles of freezing and thawing in either water or a salt solution. The durability is then defined in terms of the loss of mass, decrease in compressive strength, decrease in dynamic elastic modulus, or some combination of these. While these test methods differ from each other in detail, they all seem to provide useful information with regard to the behavior of concrete under the prescribed test conditions. They do not necessarily correlate with the behavior of concrete in the field, where the exposure conditions may be quite different from those prescribed in the various test methods.

2. *Depth of carbonation:* The depth of carbonation is generally measured by exposing a freshly fractured surface to a phenolphthalein solution; the noncarbonated portion should be pink. Though this test is very simple to perform, it may sometimes be difficult to interpret the results, particularly in areas of visible cracking.

3. *Sulfate attack:* There are currently no *standardized* test methods in Europe to determine the sulfate resistance of concrete, though there are of course a number of laboratory tests used variously in different countries. There are, however, ASTM tests for sulfate resistance. The tests all involve immersing the concrete in question in prescribed sulfate solutions, and measuring changes in length, in strength, or in dynamic modulus over the period of immersion. These tests do not correlate well with field experience, particularly if the concrete is subjected to wetting and drying cycles, and in cases in which the sulfate-bearing fluid can flow continuously past the concrete.

4. *Chloride penetration:* The rapid chloride penetration test, such as that prescribed in ASTM C1202, measures the electrical conductance, in terms of the total charge in coulombs passed over a specified time period through a concrete disc exposed to a sodium chloride solution on one side and a sodium hydroxide solution on the other. While this test is not a very good indicator of how easily chloride ions will penetrate concrete in the field, it has become quite popular as a screening test for concrete penetrability in general. This test cannot replicate the transport of chloride ions (or other aggressive ions) through concrete in practice, but is useful in a comparative manner when considering different concrete mixes. There is no particular scientific basis for this use of chloride permeability tests. However, since reducing the number of coulombs passed in a test generally involves reducing the w/b

ratio, often combined with the use of silica fume or other SCMs, a lower coulomb number does correlate empirically with increased durability to chemical attack.

5. *Alkali–aggregate reactivity:* Tests to determine the alkali–aggregate reactivity of aggregates are not concrete durability tests *per se*. They are simply designed to assess the potential of any aggregate to enter into AARs. These tests usually involve petrographic analysis to start, followed by preparation of specimens in the form of bars or prisms. These specimens either contain or are exposed to high alkali levels, and their length change over time is monitored. If the aggregates are found to be reactive, the most common solution is to move to a non-reactive aggregate, or to use cement with a very low alkali content. None of the available tests correlate perfectly with field experience.

6. *Leaching:* There are currently no tests available that can determine how readily material can be leached from concrete.

Hence, none of these tests can provide an *absolute* guarantee of future concrete behavior; they are all *relative* in nature. That is, they are not capable of providing accurate information related to deterioration rates in real-world exposures. Simply put, they cannot really

- Simulate the conditions that will exist in the field
- Deal with cracked or damaged concrete
- Provide timeline data on the state of deterioration
- Be correlated with long-term exposure
- Deal with situations in which several deterioration mechanisms are acting on the concrete at the same time
- Deal with the accelerated construction schedules that are becoming more common
- Provide clear compliance criteria for the required performance

As well, the existing tests are not robust enough for use outside of the narrowly prescribed limits laid down in the test methods themselves. What are sorely needed are tests to be used during construction to verify *in situ* properties in a timely way. Such tests must be developed in conjunction with modeling to ensure proper calibration of the test results.

While the available tests are useful in those specific cases in which we know *a priori* the destructive mechanisms that will attack the concrete (e.g., freeze–thaw, sulfates, ASR), they cannot be used as general screening tests in cases in which where this information is incomplete or uncertain. Moreover, even if test data are available, it remains very much a matter of judgment as to how the data will be used. There still need to be criteria as to the frequency of inspection and maintenance schedules, and at what point significant repairs or other rehabilitative measures must be undertaken, or when the structure should be taken out of service entirely. It is also often unclear as to *who* is responsible for deciding what level of serviceability is acceptable at any given time.

It should also be noted that virtually all of the current test methods involve the testing of *virgin* test specimens. Tests are carried out on specimens specially cast for test which are then exposed to conditions that are either arbitrarily chosen, specified in the test protocol, or which purport to simulate the environmental exposure expected in the field. That is, these specimens are then exposed to, for instance, specific cycles of freezing and thawing, specific concentrations of sulfates or other deleterious solutions, and so on. However, in practice, *the concrete will be under load* when exposed to its environment. There is now a considerable amount of literature that shows that chemical attack of concrete will be exacerbated if the concrete is under load, or if there is more than one deterioration mechanism acting simultaneously; some of this work has been summarized by Yao et al. (2012). They concluded that a combination of different processes (whether load-induced or environmental) will be much more severe than if the processes act separately. This suggests that the current test methodology tends to overestimate the durability of concrete structures in practice.

5.2 LIMIT STATE APPROACH

The move to performance-based durability design would be fully achieved by the *limit state design approach* and partially by the *partial factor design approach*. For this purpose, the design process requires quantifications of the service life of the structure/component, t_s. The service life should exceed the design service life, t_d:

$$t_s > t_d \tag{5.1}$$

Guidance for design target service life is provided in Table 5.2.

Table 5.2 Indicative values for the target service life, t_d

Class	Notional design working life (years)	Examples
1	1–5	Temporary structures
2	25	Replacement structural parts, e.g., gantry girders, bearings
3	50	Buildings and other common structures, other than those listed below
4	100 or more	Monumental buildings, and other special or important structures. Large bridges

Source: Adapted from ISO 16204, 2012, *Durability—Service Life Design of Concrete Structures*; ISO 2394, 2015, *General Principles on Reliability for Structures*, International Standards Organization, 111pp; Reproduced with the permission of the South African Bureau of Standards (SABS). The South African Bureau of Standards is the owner of the copyright therein and unauthorized use thereof is prohibited. The standard may be obtained at the following: www.store.sabs.za

The service life is the point at which the limit state has been exceeded. This condition implies that the resistance is equal to or smaller than the environmental load, as expressed by the limit state equation:

$$R(t) - S(t) \leq 0 \tag{5.2}$$

where $R(t)$ is the resistance capacity of the component at time t and $S(t)$ is the action (load) effect (mechanical, physical, or chemical) at time t.

The limit state can either be ULS or SLS as quoted from ISO 2394 (2015).

5.2.1 Ultimate limit states

These are as follows:

1. Loss of equilibrium of the structure or of a part of the structure, considered as a rigid body (e.g., overturning)
2. Attainment of the maximum resistance capacity of sections, members, or connections by rupture (in some cases affected by fatigue, corrosion, etc.) or excessive deformations
3. Transformation of the structure or part of it into a mechanism
4. Instability of the structure or part of it
5. Sudden change of the assumed structural system to a new system (e.g., snap through)

The effect of exceeding a ULS is almost always irreversible and the first time that this occurs, it causes failure.

5.2.2 Serviceability limit states

These include the following:

1. Local damage (including cracking) which may reduce the working life of the structure or affect the efficiency or appearance of structural or nonstructural elements; repeated loading may affect the local damage, for example, by fatigue.
2. Unacceptable deformations which affect the efficient use or appearance of structural or nonstructural elements or the functioning of equipment.
3. Excessive vibrations which cause discomfort to people or affect nonstructural elements or the functioning of equipment.

In the cases of permanent local damage or permanent unacceptable deformations, exceeding an SLS is irreversible and the first time that this occurs, it causes failure.

In other cases, exceeding an SLS may be reversible and then failure occurs as follows:

1. The first time the SLS is exceeded, if the exceedence is considered as acceptable
2. If the exceedence is acceptable but the time when the structure is in the undesired state is longer than specified
3. If the exceedence is acceptable but the number of times that the SLS is exceeded is larger than specified
4. If a combination of the above criteria or of some other relevant criteria occur

These cases may involve temporary local damage (e.g., temporarily wide cracks), temporary large deformations, and vibrations.

Design criteria for SLSs are generally expressed in terms of limits for acceptable deformations, accelerations, crack widths, and so on.

This approach is in essence performance based, since performance requires the structure to be able to withstand external loads. In the case of durability performance, this external load could be chemical (e.g., chloride concentration at the level of the reinforcing steel), physical (e.g., moisture content), or mechanical (e.g., internal stress developed during drying). The limit state is a function of the *environmental action* and the *action effect*. However, the time it takes until the limit state is reached (i.e., service life) depends to a large extent on the kinetics of the changes, which is controlled in many cases by *transport mechanisms* such as heat flow into the concrete, penetration of chemicals, and moisture. All of these, the *structure environment*, the *transport mechanisms*, the *environment action* and the *action effect*, have to be considered and taken into account in order to model the durability performance, based on the limit state concept, as presented in Figure 5.2.

Quantitative treatment of all of these aspects requires understanding of the processes involved and their modeling by mathematical tools. This knowledge is to a large extent dependent on advances in the science of building materials and structures in the context of the investigation and modeling of deterioration processes. The limit state approach and service life design concepts in the performance-based codes provide the framework for the application of models developed by scientific research. These models, which are quantitative mathematical descriptions of the degradation processes and their kinetics, form the basis for the calculation of changes in properties in the aggressive environment, to determine at what stage they decline below the required limit state of the structure or component, and thus determine the service life. However, the reader should keep in mind the cautions that were expressed earlier as to the "realities" inherent in these models. Extreme accuracy is simply not available using these approaches.

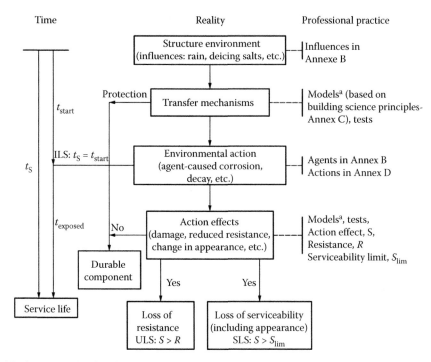

ᵃ Both conceptual and mathematical.

Figure 5.2 Limit state method for durability. *Comment*: The term transfer mechanisms in this ISO chart is synonymous with the term transport mechanisms often used in technical publications related to cement and concrete durability. *Note*: t_s, service life; t_{start}, time to initiation of degradation (years); $t_{exposed}$, time after initiation of degradation (years); ILS, initiation limit state; SLS, serviceability limit state; ULS, ultimate limit state. The annexes referred to provide a commentary with examples of how to apply these concepts in practice. (Adapted from ISO 13823, 2008, *General Principles on the Design of Structures for Durability*, ISO International Standard 13823; Reproduced with the permission of the South African Bureau of Standards [SABS].) The South African Bureau of Standards is the owner of the copyright therein and unauthorized use thereof is prohibited. The standard may be obtained at the following: www.store.sabs.za

Durability in terms of service life calculations can be determined based on the ULS approach (i.e., deterioration resulting in structural failure due to loss of resistance), or on the SLS approach (i.e., local damage, including cracking or displacements that affect the function or appearance of structural or nonstructural components), using models which predict the changes with time of the resistance capacity due to the effect of the action. Setting up the limit state equations along the principle of Equation 5.2 forms the basis for calculations for design for service life.

The action function effect, $S(t)$, in Equation 5.2 can be modeled mathematically in terms of variables based on materials science concepts. These are models which are developed to resolve and quantify processes such as corrosion of steel due to carbonation and chloride ingress and deterioration of the concrete due to frost action and chemical attack. Demonstration of the application of such models is provided in Section 5.5, based on examples outlined in various codes and studies.

5.3 PROBABILISTIC LIMIT STATE APPROACH

Reference to this approach has already been made in the preceding section. Durability calculations and design are usually based on models that quantify the penetration of deleterious materials into the concrete. In the case of corrosion of steel in concrete, the limit state is defined in terms of the buildup of a critical content of the penetrating material at the level of the steel, which causes the limit state distress, either depassivation of the steel, or cracking and spalling, as shown schematically in Figure 5.3.

The deterministic approach is based on the time to reach the critical level of concentration, as shown in Figure 5.4a for a chloride penetration profile after 50 years of exposure. The critical buildup is at a depth of 34.1 mm, implying

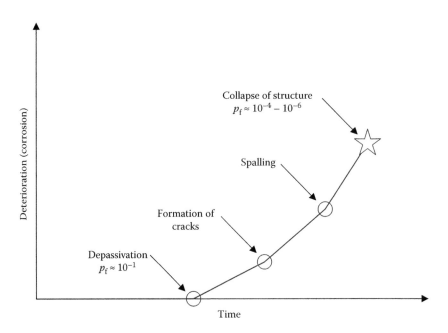

Figure 5.3 Limit states and related reliability levels. (After Helland, S., 2013, Design for service life: Implementation of FIB Model Code 2010 rules in the operational code ISO 16204, *Structural Concrete*, 14(1), 10–18.)

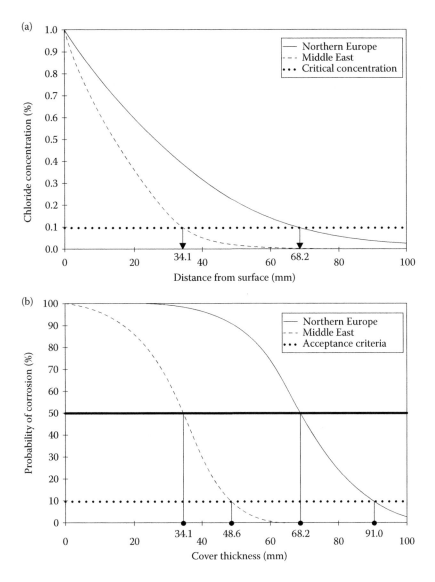

Figure 5.4 Deterministic approach (a) and probabilistic approach (b) for determining the concrete cover depth to achieve 50 years durability of the reinforcing steel in a chloride environment. The limit state for depassivation is taken in this example as 0.1% of chlorides by weight of cement and the probabilistic calculation is based on specifying that the risk of depassivation within the designed service life of 50 years will be smaller than 10%. *Note*: The limit state for depassivation used in Figure 5.4 (0.1% by weight of cement) is lower than the commonly accepted value, which is usually between 0.20% and 0.4% by weight of cement. (After fib, 1999, *Textbook on Behavior, Design and Performance—Updated Knowledge of the CEB/FIP Model Code 1990*, Vol. 3, Bulletin 3.)

that a concrete cover of that value is required. However, since the calculation is based on average values of the penetration parameters and depth of cover, design based on this outcome implies that it will provide only a 50% probability of achieving the required corrosion free service life. However, from an engineering point of view, the requirement is that the probability of corrosion during that period should be small. If, for example, a 10% probability of reaching the critical level is specified, then the concrete cover should be greater to an extent quantified by statistical considerations (Figure 5.4b).

Thus, statistical considerations are very important for assessing durability performance and design for durability. This is as true for the environmental effects that cause the actions, as for the characteristics of the structure or component that resist such actions, for example, the concrete cover over the reinforcing steel. With this in mind, the limit state equation (Equation 5.2) can be described in statistical terms of the probability target, P_f, as shown schematically in Figure 5.5.

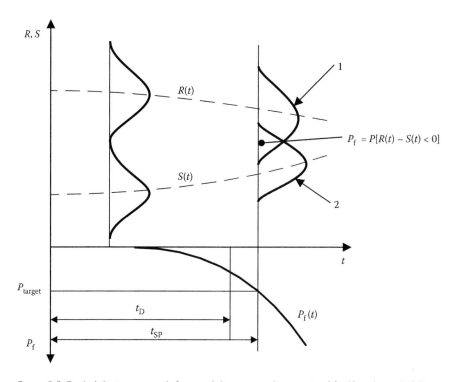

Figure 5.5 Probabilistic approach for modeling to predict service life. Key: 1, probability density function of $R(t)$; 2, probability density function of $S(t)$. (Adapted from ISO 13823, 2008, *General Principles on the Design of Structures for Durability*, ISO International Standard 13823; Reproduced with the permission of the South African Bureau of Standards [SABS]. The South African Bureau of Standards is the owner of the copyright therein and unauthorized use thereof is prohibited. The standard may be obtained at the following: www.store.sabs.za)

The mathematical expression to describe the statistical concepts in Figure 5.5 for the various durability formats are outlined in Equation 5.3 in terms of target failure probability, p_f:

$$P_f(t) = P[R(t) - S(t) < 0] < p_{f(\text{target-ULS})} \tag{5.3}$$

The choice of value for the target probability for failure, p_f, is an issue of judgment, which depends on considerations of the consequences of the failure as well as agreement between the client and the appropriate authority. ISO 13823 suggests four categories depending on the consequences of failure (Table 5.3). Examples of suggested p_f values are provided in Figure 5.3.

The *fib* Model Code for Service Life Design and ISO 13823 suggest three categories, which are based on the consequences of failures and malfunctions, as outlined in Table 5.4.

Table 5.3 Categories of component failure based upon consequences

Category	Consequences of failure	Examples
1	Minor and repairable damage, no injuries to people	Components where replacement after failure is planned for, or where other reasons for replacement are more relevant, like coatings or sealants
2	Minor injuries or little disruption of the use and occupancy of the structure, including components that protect other components essential for the function of the assembly	Replaceable but important components for the function of the structure, such as installations for heating, lighting, and ventilation, or windows, whose replacement is planned before failure
3	Nonserious injuries or moderate economic, social, or environmental consequences	Nonheavy, nonstructural components of the facility requiring major repair work if they fail, such as plumbing, or components/systems whose replacement is planned before failure, such as structural bearings and railings or cladding
4	Loss of human life or serious injuries, or considerable economic, social, or environmental consequences	Structural components that are parts of the primary or secondary load-carrying system, emergency exits, or components causing major damage if they fail (e.g., heavy parts of the envelope, prefabricated wall elements, heavy inner walls)

Source: Adapted from ISO 13823, 2008, *General Principles on the Design of Structures for Durability*, ISO International Standard 13823; Reproduced with the permission of the South African Bureau of Standards (SABS). The South African Bureau of Standards is the owner of the copyright therein and unauthorized use thereof is prohibited. The standard may be obtained at the following: www.store.sabs.za

Table 5.4 Classification of consequences of failures in terms of consequence classes (CC) suggested in the *fib* Model Code for Service Life Design (2006)

Consequences classes	Description	Examples of buildings and civil engineering works
CC3	High consequences for loss of human life or economic, social, or environmental consequences—very great	Grandstands, public buildings where consequences of failure are high (e.g., concert hall)
CC2	Normal consequences for loss of human life and economic or environmental consequences—considerable	Residential and office buildings, public buildings where consequences of failure are medium (e.g., office buildings)
CC1	Low consequences for loss of human life and economic or environmental consequences are small or negligible	Agricultural buildings where people do not normally enter (e.g., storage buildings, green houses)

Source: Adapted from *fib* Model Code for Service Life Design (*fib* MC-SLD), 2006, *fib* Bulletin no. 34, 2006, *fib* secretariat, Case postale 88, CH-1015, Lausanne, Switzerland; EN 1990 2002.

Reliability classes based on the consequence classes in Table 5.4 can be quantified in terms of the target probability of failure, p_f. The probability of failure, p_f, can be substituted by a reliability index, β, which is defined as

$$\beta = -\Phi^{-1}(p_f)$$ (5.4)

where Φ^{-1} is the inverse standardized normal distribution.

The relationship between β and p_f is presented in Table 5.5.

Codes and standards often recommend values for β and p_f based on general considerations of the consequences of failure and costs of safety measures. Table 5.6 (ISO 2394, 1998) presents the recommendations of ISO 2394, which considers the costs of safety measures and the consequences of failure. Table 5.7 provides recommendations for reliability indices based on the deterioration processes in different environmental conditions. This table is derived from the *fib* Model Code for Service Life Design (2006) for the exposure classes specified in Eurocode 2. EN 1990 (Eurocode)

Table 5.5 Relationship between β and p_f

p_f	10^{-1}	10^{-2}	10^{-3}	10^{-4}	10^{-5}	10^{-6}	10^{-7}
β	1.3	2.3	3.1	3.7	4.2	4.7	5.2

Source: Adapted from ISO 2394, 2015, *General Principles on Reliability for Structures*, International Standards Organization, 111pp; Reproduced with the permission of the South African Bureau of Standards (SABS). The South African Bureau of Standards is the owner of the copyright therein and unauthorized use thereof is prohibited. The standard may be obtained at the following: www.store.sabs.za

Table 5.6 Target β values recommended by ISO 2394 1998

Relative costs of safety measures	Consequences of failure			
	Small	Some	Moderate	Great
High	0	A 1.5	2.3	B 3.1
Moderate	1.3	2.3	3.1	C 3.8
Low	2.3	3.1	3.8	4.3

Source: Reproduced with the permission of the South African Bureau of Standards (SABS). The South African Bureau of Standards is the owner of the copyright therein and unauthorized use thereof is prohibited. The standard may be obtained at the following: www.store.sabs.za

Note: A: for SLSs, use β = 0 for reversible and β = 1.5 for irreversible limit states; B: for fatigue limit states, use β = 2.3 to 3.1, depending on the possibility of inspection; C: for ULSs design, use the safety classes β = 3.1, 3.8, and 4.3. These numbers have been derived with the assumption of lognormal or Weibull models for resistance, Gaussian models for permanent loads, and Gumbel extreme value models for time-varying loads and with the design value method according to E.6.2. It is important that the same assumptions (or assumptions close to them) are used if the values given in Table E.2 are applied for probabilistic calculations. Finally, it should be stressed that a β-value and the corresponding failure probability are formal or notional numbers, intended primarily as a tool for developing consistent design rules, rather than giving a description of the structural failure frequency.

Table 5.7 Recommended minimum values of reliability index β and failure probability (p_f) intended for the design of service life

Exposure class-Eurocode 2	Description	Reliability class	SLS depassivation β (p_f)	ULS collapse β (p_f)
XC	Carbonation	RC1	1.3 ($\approx 10^{-1}$)	3.7 ($\approx 10^{-4}$)
		RC2	1.3 ($\approx 10^{-1}$)	4.2 ($\approx 10^{-5}$)
		RC3	1.3 ($\approx 10^{-1}$)	4.4 ($\approx 10^{-6}$)
XD	Deicing salts	RC1	1.3 ($\approx 10^{-1}$)	3.7 ($\approx 10^{-4}$)
		RC2	1.3 ($\approx 10^{-1}$)	4.2 ($\approx 10^{-3}$)
		RC3	1.3 ($\approx 10^{-1}$)	4.4 ($\approx 10^{-6}$)
XS	Seawater	RC1	1.3 ($\approx 10^{-1}$)	3.7 ($\approx 10^{-4}$)
		RC2	1.3 ($\approx 10^{-1}$)	4.2 ($\approx 10^{-3}$)
		RC3	1.3 ($\approx 10^{-1}$)	4.4 ($\approx 10^{-6}$)

Source: Adapted from *fib* Model Code for Service Life Design (*fib* MC-SLD), 2006, *fib* Bulletin no. 34, 2006, *fib* secretariat, Case postale 88, CH-1015, Lausanne, Switzerland.

Table 5.8 Recommended minimum values for reliability
index β for 50-year reference period (ULS)

Reliability class	Minimum reliability values, β (p_f)
RC3	4.3 ($\approx 1 \times 10^{-3}$)
RC2	3.8 ($\approx 7 \times 10^{-3}$)
RC1	3.3 ($\approx 48 \times 10^{-3}$)

Source: Adapted from Table B2 in EN 1990:2002 (the values in
parenthesis were added and they represent the
approximate corresponding p_f values).

Note: The three reliability classes are associated with the three
consequence classes described in Table 5.3.

considers the consequences classes CC1, CC2, and CC3 in Table 5.4 and
associates them with three reliability classes RC1, RC2, and RC3 to recommend minimum values for reliability index β (Table 5.8).

5.4 PARTIAL FACTOR DESIGN APPROACH

The models used for the partial factor design are essentially the same as
those used for the probabilistic approach, except that the treatment of the
probabilistic uncertainties are expressed in terms of partial factors, considering uncertainties in the values and in the models. These include the possibility for unfavorable deviations of action values, materials, and product
properties from the representative values, as well as model and dimensional
uncertainties. The numerical values for the partial factors should be determined on the basis of statistical evaluations of experimental data and field
observations and should also consider calibration to long-term experience.

The main basic variables to consider are those which are of prime
importance for the design results. They include the representative values for
actions (i.e., the value used for the verification of a limit state, which could
either be the characteristic value or the accompanying value of a variable
action which may be the combination value, the frequent value, or a quasi-
permanent value), materials properties, and geometrical parameters as well
as the model in which the design values are incorporated to validate the
limit state so that

$$R_d > S_d \qquad (5.5)$$

where R_d is design resistance and S_d is the design actions.

Formulations of the partial factors are outlined below:

Design values for actions:

$$F_d = \gamma_f \cdot F_{rep} \qquad (5.6)$$

where F_{rep} is the representative value of the action and γ_f is the partial factor for action.

Design value for material properties:

$$f_d = \frac{f_k}{\gamma_m} \tag{5.7}$$

where f_k is the characteristic value of material property, where the characteristic value is represented by *a priori* specified fractile of the statistical distribution of the material property, and γ_m is the partial factor for material property.

Design value for geometrical quantities:

$$a_d = a_k \pm \Delta a \tag{5.8}$$

where a_k is the characteristic value of geometrical quantity and Δa is the geometrical deviation, which includes tolerance specification.

Model uncertainties:

Design values may take into account the uncertainties in the design model of resistance as well as in the property value:

$$R_d = \frac{R_k}{\gamma_M} = \frac{R_k}{(\gamma_m \cdot \gamma_{Rd})} = \frac{R_k}{\gamma_M} \tag{5.9}$$

where R_k is the characteristic value of resistance, γ_m is the partial factor for material property, γ_{Rd} is the partial factor associated with uncertainty of the resistance model plus geometrical deviations if these are not modeled explicitly, and $\gamma_M = \gamma_m \cdot \gamma_{Rd}$ is the partial factor accounting for material property and uncertainties in model and dimensional variations.

Service life:

Similar concepts of partial factors can be defined for design of service life to meet the specified design service life, t_d:

$$\frac{t_{sk}}{\gamma_S} \geq t_d \text{ and } \gamma_S \geq 1.0 \tag{5.10}$$

where t_{sk} is the characteristic value of calculated service life and γ_S is the partial factor.

5.5 APPLICATION OF THE LIMIT STATE DESIGN METHODS

The limit state design methods can provide the quantitative basis for the performance approach. In order to apply these methods, especially those taking into consideration probabilistic considerations, such as the full probabilistic and the partial factor methods, there is a need for (i) relevant models, (ii) parameters that can be used with the models, and (iii) test methods that can be applied to determine material properties and assess the effect of actions. Intensive effort in all of these aspects has taken place in recent decades.

Although these models and input variables can be incorporated into the probabilistic limit state design methods, there are as yet no models and input data which have national or international consensus. The Japanese Standard Specification for Concrete Structures—2007: Design (JSCE, 2007) tends to provide details on such data, whereas ISO 16204 (2012) and *fib* Model Code 2010 do not make recommendations for input data, suggesting that such data will be derived from the literature or existing structures where the concrete composition, execution, and exposure conditions are similar to those expected for the new structure. Examples and sources for data are referred to in commentaries in the codes and examples in the standards. Systematic compilations of data to back these standards and codes are available in *fib* Model Code for Service Life Design (2006). They are based on several European projects, which started with the DuraCrete network (DuraCrete, 2000) and continued with DARTS (2004a,b) as well as the work done jointly by various organizations in the United States, which resulted in LIFE-365, under the umbrella of ACI (ACI 365.1R-00, 2000; LIFE-365, 2012).

These developments are reviewed in this chapter to demonstrate the advances that have taken place so far, and their potential to become common and acceptable design tools. The modeling and quantification approach taken is to consider each deterioration process on its own and to provide the quantitative tools and data. This is consistent with the guidance already given in some codes and standards that classify the environmental conditions in terms of the degradation processes that they may evoke, as is the case in EN 206 and its application in Eurocode 2.

The data that need to be provided for these purposes are not just the values of the characteristic input parameters for the models, but also their standard deviations (s.d.). For example, the concrete cover is specified in terms of a mean value. The s.d. values suggested are 8–10 mm where there are no particular construction requirements, and 6 mm when special construction requirements are specified. Normal distributions are commonly applied, but for small cover, there is a need to use distributions which would eliminate negative cover values.

5.5.1   Limit states for corrosion of steel

The limit state can be defined as depassivation of the reinforcing steel by either carbonation or chloride ingress, or the onset of actual damage after corrosion propagation has taken place following the depassivation.

The definitions of limit state, whether depassivation or different indicators of damage, are shown schematically in Figure 5.6. The limiting states in Figure 5.6 can be associated with reliability levels, as outlined in Figures 5.3 and 5.4b.

When the limit state is defined as depassivation of the reinforcing steel, the criterion for durability is the difference between the concrete cover, c, and the depth of penetration of carbonation or a critical front of critical chloride content, $x_c(t)$, or simply the depth of penetration in relation to the concrete cover. A demonstration of this durability state is provided in Figure 5.7 for the case of carbonation, based on ISO 13823-2008. The service life can be calculated from the probability consideration of ULS format (Equation 5.11):

$$P_f(t) = P[x_c(t) - c < 0] < P_{f(\text{target})} \tag{5.11}$$

The *fib* Model Code 2010 provides guidelines for the use of the partial factor approach, demonstrating it for the case of carbonation. In this approach, the design cover, c_d, is defined in terms of the nominal value, c_{nom}, and deviations from it, Δc, which can be the result of several causes, such as inaccuracy in the construction:

$$c_d = c_{\text{nom}} \pm \Delta c \tag{5.12}$$

The limit state is defined as

$$c_d - x_{cd}(t_{SL}) > 0 \tag{5.13}$$

where $x_{cd}(t_{SL})$ is the design value of carbonation depth at time t_{SL} and t_{SL} is the service life time.

The depth units are in millimeters.

The design value of the penetration depth, $x_{cd}(t_{SL})$, is calculated as

$$x_{cd}(t_{SL}) = x_{c,c}(t_{SL}) \cdot \gamma_f \tag{5.14}$$

where $x_{c,c}(t_{SL})$ is the characteristic value of carbonation depth at the service life time, t_{SL}, and γ_f is the partial safety factor for the carbonation depth.

Note: The *characteristic value* is the value of a material or product property having a prescribed probability of not being attained in a hypothetical unlimited test series. This value generally corresponds to a specified fractile of the assumed statistical distribution of the particular property of the material or product. A nominal value is used as the characteristic value in some circumstances (*fib* Bulletin 34, 2006).

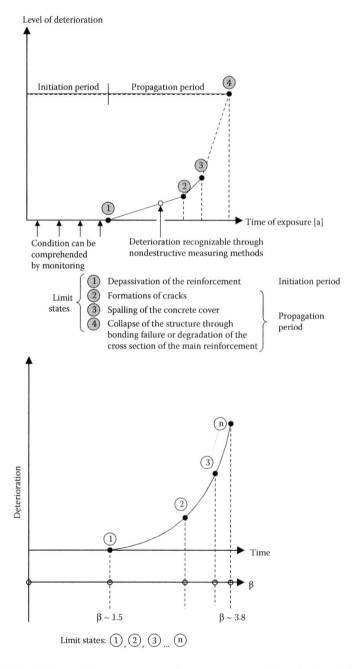

Level of deterioration

Initiation period | Propagation period

Time of exposure [a]

Condition can be comprehended by monitoring

Deterioration recognizable through nondestructive measuring methods

Limit states
1. Depassivation of the reinforcement — Initiation period
2. Formations of cracks
3. Spalling of the concrete cover — Propagation period
4. Collapse of the structure through bonding failure or degradation of the cross section of the main reinforcement

Deterioration

Time

β

$\beta \sim 1.5$ $\beta \sim 3.8$

Limit states: (1), (2), (3) ... (n)

Figure 5.6 Definitions of limit states according to corrosion of steel-induced damage. (Adapted from *fib* Model Code for Service Life Design (*fib* MC-SLD), 2006, *fib* Bulletin no. 34, 2006, *fib* secretariat, Case postale 88, CH-1015, Lausanne, Switzerland.)

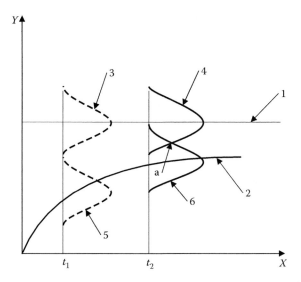

Figure 5.7 Probability of failure due to carbonation. Key: X, time, t, expressed in years; Y, cover depth, S_{lim} (for curves 1, 3, and 4), or the depth of the carbonated zone, $S(t)$ (for curves 2, 5, and 6); 1, curve of the cover depth, S_{lim}; 2, curve of the depth of the carbonated zone, $S(t)$; 3, curve of the distribution of cover depths, S_{lim}, at time t_1; 4, curve of the distribution of cover depths, S_{lim}, at time t_2; 5, curve of the distribution of the depths of the carbonated zone, $S(t)$, at time t_1; 6, curve of the distribution of the depths of the carbonated zone, $S(t)$, at time t_2; a, the overlap of the distributions shown as curves 4 and 6 indicates the possibility of a failure. (Adapted from ISO 13823, 2008, *General Principles on the Design of Structures for Durability*, ISO International Standard 13823; Reproduced with the permission of the South African Bureau of Standards [SABS].) The South African Bureau of Standards is the owner of the copyright therein and unauthorized use thereof is prohibited. The standard may be obtained at the following: www.store.sabs.za

The equations for the full probabilistic and partial factor approach have been demonstrated here for the depth of carbonation. However, they are formulated in general terms, which is also applicable for design calculations of other durability problems associated with the penetration of deleterious substances, such as chlorides that depassivate steel when reaching a critical concentration level at the steel surface, and chemicals that lead to a chemical attack on the concrete. Applications for carbonation-induced corrosion (Section 5.5.1.1) and chloride-induced corrosion (Section 5.5.1.2) follow.

5.5.1.1 Carbonation-induced corrosion

A range of models has been developed for the depth of carbonation over time. Most of them result in a linear relation between carbonation depth and square root of time:

$$x_c(t) = K \cdot t^{1/2} \tag{5.15}$$

The value of K is a function of numerous parameters, including the concrete material itself, construction practices and environmental conditions. Relations of this kind have been developed by numerous investigators, based on laboratory testing in normal and accelerated conditions. Similar relations that have been developed and verified for local conditions can be used for the analysis.

The ISO13823-2008 standard presents an example of a calculation based on the carbonation equation developed by (Kasami et al., 1986):

$$X_c(t) = \alpha \cdot \beta \cdot \gamma \cdot t^{1/2} \tag{5.16}$$

where $x_c(t)$ is the carbonation depth at time t (mm); α is the coefficient depending on local environmental actions, such as concentrations of CO_2, temperature, and humidity; β is the coefficient depending on finishing material on the concrete surface; and γ is the coefficient depending on the quality of concrete.

The limit state equation is given by the difference between the cover depth and the depth of carbonation, and its probabilistic expression (Equation 5.17) for normal distribution is as follows:

$$f[S_{\lim} - S(t)] = \frac{1}{\sqrt{2\pi[\bar{S}(t)^2 V^2 + \sigma^2]}} \cdot \exp\left\{ \frac{-\{[S_{\lim} - S(t)] - [\bar{S}_{\lim} - \bar{S}(t)]\}^2}{2[\bar{S}(t)^2 V^2 + \sigma^2]} \right\} \tag{5.17}$$

where $\bar{S}(t)$ is the mean value of the depth of carbonation, V is the coefficient of variation of the depth of carbonation, $\bar{S}_{\lim}$ is the mean value of the cover depth, and σ is the s.d. of the cover depth.

A demonstration of the calculation outcome of the probability for corrosion initiation based on this equation is shown in Figure 5.8, using the values for Equations 5.16 and 5.17 as follows:

α—1.0 (outdoor concrete)
β—1.0 (no finishing of concrete surface)
γ—a function of w/c ratio, using that in the paper by Kasami et al. (1986)
V—0.4
σ—15 mm

ISO 16204:2012 and *fib* Model Code 2010 suggest a different form for the coefficients preceding the square root of time:

$$x_c(t) = k \cdot W(t) \cdot t^{1/2} \tag{5.18}$$

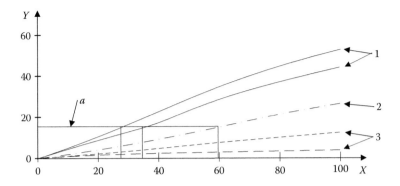

Figure 5.8 Relation between probability of corrosion and predicted service life. Key: X, time, t, expressed in years; Y, probability of corrosion initiation, P, expressed in percentage; 1, high w/c; 2, medium w/c; 3, low w/c; a, $P_{target} = 15\%$. (Adapted from ISO 13823, 2008, *General Principles on the Design of Structures for Durability*, ISO International Standard 13823; Kasami, H. et al. 1986, Carbonation of concrete and corrosion of reinforcement in reinforced concrete, *Proceedings of First Australia–Japan Coordinated Workshop on Durability of Reinforced Concrete*, Building Research Institute, Japan, C4, October, pp. 1–13; Reproduced with the permission of the South African Bureau of Standards [SABS].) The South African Bureau of Standards is the owner of the copyright therein and unauthorized use thereof is prohibited. The standard may be obtained at the following: www.store.sabs.za

where k is the factor reflecting the concrete properties in reference conditions and $W(t)$ is the weather function taking into account the mesoclimatic conditions due to wetting events of the concrete surface.

In-depth information on calculations and data required for this model are provided in the *fib* Model Code for Service Life Design (2006), based on knowledge developed in the DuraCrete project, which is used for $x_c(t)$ in Equation 5.15. It is based on concepts of diffusion assuming that the CO_2 diffusion coefficient through concrete is a material parameter, although it may depend on numerous factors during the service life:

$$X_c(t) = [2 \cdot k_e \cdot k_c \cdot R_{ANC,0}^{-1} \cdot C_s]^{1/2} \cdot W(t) \cdot (t)^{1/2} \tag{5.19}$$

where $R_{ANC,0}^{-1}$ is the inverse of the resistance to carbonation in natural conditions; if accelerated carbonation tests are used, then this term is replaced by $(k_t \cdot R_{ACC,0}^{-1} + \varepsilon_t)$, which transforms the accelerated test results to inputs for carbonation in natural environment:

$$X_c(t) = \left[2 \cdot k_e \cdot k_c \cdot \left(k_t \cdot R_{ACC,0}^{-1} + \varepsilon_t\right) \cdot C_s\right]^{1/2} \cdot W(t) \cdot (t)^{1/2} \tag{5.20}$$

where $x_c(t)$ is the carbonation depth at the time t (mm), t is time (s), k_e is the environmental function, k_c is the execution transfer parameter, k_t is the regression parameter, $R_{ACC,0}^{-1}$ is the inverse effective carbonation resistance of concrete determined by accelerated test in CO_2 concentration of 2% [(mm²/s)/(kgCO₂/m³)], ε_t is the error term, C_s is the CO_2 concentration (kgCO₂/m³), and $W(t)$ is the weather function.

The values and calculations of these parameters are outlined below:

1. *Environmental function, k_e.* This function takes into account the effect of humidity on the diffusion coefficient relative to reference conditions (lab conditions) of 20°C/65%RH:

$$K_e = \left\{ \left[1 - \left(\frac{RH_{real}}{100} \right)^{f_e} \right] \middle/ \left[1 - \left(\frac{RH_{ref}}{100} \right)^{f_e} \right] \right\}^{g_e} \tag{5.21}$$

where RH_{real} is the relative humidity of the carbonated layer (%; can be estimated by local weather monitoring and applying adequate distribution function to describe it); RH_{ref} is the reference relative humidity (65%); f_e is the exponent, reported empirically to be 2.5; and g_e is the exponent, reported empirically to be 5.0.

2. *Execution transfer parameter, k_c.* This parameter accounts mainly for the influence of curing and it can be described as

$$K_c = \left(\frac{t_c}{7} \right)^{b_c} \tag{5.22}$$

where t_c is the duration of curing and b_c is the exponent of regression.

3. *Inverse carbonation resistance, $R_{ACC,0}^{-1}$.* This is a characteristic of the concrete resistance to carbonation, which is evaluated by accelerated testing. The s.d. of this property was shown to be 69% of the average $R_{ACC,0}^{-1}$ value. Guidelines in the case of lack of data are provided in Table 5.9.

   The parameters k_t and ε_t in Equation 5.20 are test method factors for the accelerated carbonation test, being 1.25 (s.d. 0.35) and 315.5 (s.d. 48), respectively.

4. *CO_2 concentration in the atmosphere, $C_{s,atm}$.* An average value of 0.00082 kgCO₂/m³ is recommended by the *fib* Model Code for Service Life Design (2006), with 0.0001 s.d.

5. *Weather function, $W(t)$.* Taking into account wetting conditions,

$$W(t) = \left(\frac{t_o}{t} \right)^{((p_{sr}^{-ToW})^{b_W})/2} = \left(\frac{t_o}{t} \right)^{W} \tag{5.23}$$

Table 5.9 Guidelines for values for inverse carbonation resistance in accelerated conditions, as a function of w/c ratio and type of binder

Cement type	$R_{ACC,0}^{-1}$ [(mm²/s)/(kgCO₂/m³)], at equivalent w/c ratios of					
	0.35	0.40	0.45	0.50	0.55	0.60
CEM I 42.5R	–	3.1	5.2	6.8	9.8	13.4
CEM I 42.5R + FA (k = 0.5)	–	0.3	1.9	2.4	6.5	8.3
CEM I 42.5R + SF (k = 2.0)	3.5	5.5	–	–	16.5	–
CEM III/B 42.5	–	8.3	16.9	26.6	44.3	80.0

Source: Adapted from *fib* Model Code for Service Life Design (*fib* MC-SLD), 2006, *fib* Bulletin no. 34, 2006, *fib* secretariat, Case postale 88, CH-1015, Lausanne, Switzerland.

where t_0 is the time of reference, W (weather function) = $(t_0^{-ToW})^{b_W}/2$, ToW is the time of wetness = (days in a year with rainfall /365, p_{sr} is the probability of driving rain, and b_W is the regression.

al values for these parameters are provided in (*fib*, 2006)

tive for carbonation calculation, which is provided in the d Specification for Concrete Structures—2007: Design, are root model for carbonation depth, and in its com- estimates for the carbonation rate, α_p in units of mm/ ion of w/b ratio, based on experimental data:

$$\alpha_p = -3.57 + 9.0\left(\frac{w}{b}\right)(mm/(year)^{1/2}) \qquad (5.24)$$

where w/b is the effective water to binder ratio:

$$\frac{w}{b} = \frac{w}{(C + kA)} \qquad (5.25)$$

where C is the cement content, A is mineral additive content, and k is the efficiency coefficient of the mineral additive. The k values are 0 and 0.7 for fly ash and slag, respectively.

For design purposes, the Japanese specifications take the approach of partial factors, which are combined into the square root modeling of the carbonation.

The characteristic rate of carbonation, α_k, is derived from the estimated one, α_p:

$$\alpha_p \cdot \gamma_p \leq \alpha_k \qquad (5.26)$$

where γ_p is a factor taking into account the inaccuracy in determining of α_p, being between 1.0 and 1.3.

The square root model of carbonation depth recommended by the Japanese specification taking into account partial factors is

$$y_d = \gamma_{cb} \cdot \alpha_d \cdot (t)^{1/2} \tag{5.27}$$

where the parameters are already the design ones, for example, taking probability considerations by partial factors; y_d is the design value for carbonation depth; t is the design service life; γ_{cb} is the partial factor to account for the variation in the design value of the carbonation depth, normally taken as 1.15 (in the case of high fluidity concrete, it may be reduced to 1.1); and α_d is the design carbonation rate (mm/(year)$^{1/2}$).

The design carbonation rate, α_k, is a function of the characteristic carbonation rate, environmental conditions, and quality of the concrete in the structure:

$$\alpha_d = \alpha_k \cdot \beta_e \cdot \gamma_c \tag{5.28}$$

where α_k is the characteristic value of carbonation rate (mm/(year)$^{1/2}$); β_e is the factor indicating the degree of environmental influence, taken usually as 1.6 in relatively dry environment and 1.0 in relatively wet environment; and γ_c is the factor to account for the materials properties in the concrete, usually taken as 1.0 and 1.3 for the upper portions of the structure; if there are no differences in the quality of the concrete in structure and that of laboratory cured specimens, the value of 1.0 may be adopted for the whole structure.

The calculation of the required concrete cover, c, based on the design value for carbonation depth, y_d, follows these conditions for carbonation resistance verification:

$$\gamma_i \left(\frac{y_d}{y_{lim}} \right) < 1.0 \tag{5.29}$$

where γ_i is the factor representing the importance of the structure, taken usually as 1.0 and may be increased to 1.1 for important structures; and y_{lim} is the critical carbonation depth for steel corrosion, obtained from the following relations:

$$y_{lim} = c_d - c_k \tag{5.30}$$

where c_d is the design value for concrete cover used for durability verification (mm), taking into account the concrete cover, c, and construction error, Δc_e

$$c_d = c - \Delta c_e \qquad (5.31)$$

and c_k is the remaining noncarbonated cover thickness (mm), taken as 10 mm for structures in normal environments and between 10 and 25 mm for structures located in chloride-rich environments.

A sample design provided in the commentary to the Japanese Standard is shown in Figure 5.9, presenting c_d as a function of the concrete quality quantified in terms of its w/c ratio. The partial factors in this example are as follows:

$\gamma_{cb} = 1.15$
$\gamma_c = 1.0$
$\gamma_i = 1.0$
$\gamma_p = 1.0$

The environmental conditions are dry ($\beta_e = 1.6$) and the specified service life, t, is 100 years.

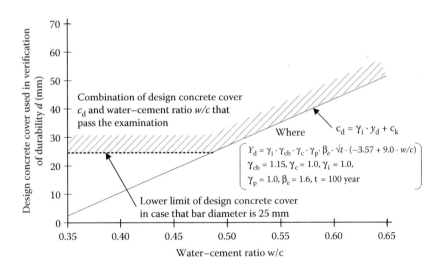

Figure 5.9 Sample relation between the thickness of carbonated concrete and water–cement ratio. (Adapted from Japanese Standard Specifications for Concrete Structures, 2007, Design, JSCE Guidelines for Concrete No. 15, Japan Society of Civil Engineers.)

5.5.1.2 Chloride-induced corrosion

The limit states for chloride-induced corrosion could either be the initiation of corrosion or an advanced state of propagation where damage has been inflicted by concrete spalling and loss of steel cross section. In numerous life-cycle estimates, the limit state of corrosion initiation is chosen as the limit state for the design of service life. This limit state is governed by the buildup of critical chloride content at the level of the steel in the concrete, leading to its depassivation. Chloride penetration into concrete has usually been modeled as a diffusion process, and accordingly the time to reach this critical level can be described in terms of Fick's second law, using Equation 5.32:

$$C_{crit} = C(x = a; t) = Co + (C_s, \Delta x - C_0)\left[1 - \text{erf}\frac{a - \Delta x}{2(D_{app,C}(t))^{1/2}}\right] \qquad (5.32)$$

where $C(x = a,t)$ is the concentration of chlorides at the depth "a" of the reinforcement at time t; C_{crit} is the critical chloride content (%wt.); $C(x,t)$ is the content of chlorides in the concrete at a depth x (structure surface: $x = 0$ mm) and at time t (%wt.); C_0 is the initial chloride content of the concrete (%wt.); Δx is the depth of the convection zone (external layer in concrete cover, up to which the process of chloride penetration differs from Fick's second law of diffusion and is governed by capillary processes) (mm); $C_{S,\Delta x}$ is the chloride content at a depth Δx at a certain point of time t (%wt.); x is the depth with a corresponding content of chlorides $C(x,t)$ (mm); a is the concrete cover (mm); $D_{app,C}$ is the apparent coefficient of chloride diffusion through concrete (mm^2/s); t is time (s); and erf is error function.

The above equation or a similar one is used by ISO 16204:2012, *fib* Model Code 2010, the Japanese Specifications for Concrete Structures—2007: Design, and ACI 365.1R-00 (Service-Life Prediction-State of the Art Report), without considering the convection zone (i.e., Δx is assumed to be 0) implying that the only mechanism for penetration is diffusion with a constant apparent diffusion coefficient, D_{app}. The apparent diffusion coefficient depends on several factors and is essentially an equivalent value for a diffusion coefficient which changes over time and can be quantified in terms of the relation below:

$$D_{app}(t) = D_{app}(t_0) \cdot \left(\frac{t_0}{t}\right)^{\alpha} \qquad (5.33)$$

where $D_{app}(t_0)$ is the apparent diffusion coefficient at reference time t_0 and α is the aging factor representing the decrease of the diffusion coefficient over time, depending on the type of binder and environmental conditions, in the range of 0.2–0.8.

The nature of outputs obtained in such models can be clearly seen in Table 5.10 from ACI 365.1R-00, in which the service life (time to initiation of corrosion) was calculated as a function of the quality of the concrete (in terms of its diffusion coefficient) and the concrete cover thickness. Table 5.10 shows clearly that concrete quality is far more critical than cover in achieving durability, although cover is also important.

The diffusion approach is in many instances oversimplified, since it assumes that (i) diffusion is the only penetration mechanism, (ii) a constant apparent diffusion coefficient can be taken for the solution of the diffusion equation, and (iii) the concrete is saturated.

Some of these simplifications are addressed in European (DuraCrete, 2000; DARTS, 2004a,b; *fib*, 2006) and North American LIFE-365 (LIFE-365, 2012) approaches. LIFE-365 has chosen to implement Fick's law in a finite difference model to enable changes in the exposure conditions and concrete properties over time to be accounted for, following a similar treatment by Boddy et al. (1999).

Since the concrete is rarely saturated in practice, ingress mechanisms other than diffusion might take place, especially capillary absorption which causes the buildup of the surface chloride concentration. Both the European (DuraCrete, 2000) and the North American (LIFE-365, 2012) models take this into account and modify the diffusion equation slightly.

The European model has considered the convection mechanism for modeling the buildup of chlorides on the concrete surface. It expresses it in terms of a convection layer in Equation 5.32, in which the chloride content is taken as the chloride content at the perimeter of the convection zone. These concepts are summarized in Figure 5.10 describing the moisture content and chloride concentration in the splash zone, where the surface concrete is not saturated and the chloride content is built-up by convection-capillary processes.

In the long run, one may expect improvements in these models to account for the complex ingress mechanisms. Thus, combined diffusion

Table 5.10 Effect of cover quality on time to initiation of corrosion of reinforced concrete

	Concrete quality (chloride diffusion coefficient, D, m^2/s)		
Cover (mm)	Low (5×10^{-11})	Medium (5×10^{-12}) Time (years)	High (5×10^{-13})
25	0.56	5.6	56
50	2.3	23	230
75	5.0	50	500

Source: Adapted from data in ACI 365.1R-00, 2000, Service Life Prediction.

Note: The critical chloride content is 0.4% by weight of the concrete; $C_{crit}/C_0 = 0.5$.

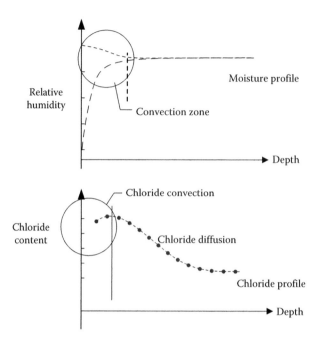

Figure 5.10 Schematic description of the moisture and chloride profiles of the surface of concrete in the splash zone. (Adapted from LIFECON, 2003, Service life models instructions on methodology and application of models for the prediction of the residual service life for classified environmental loads and types of structures in Europe.)

and convective mechanisms have to be quantified in the models (Saetta et al., 1993; Martín-Pérez et al., 1998). However, this may result in a complex quantitative treatment which could not be readily applied for engineering purposes, as there would be a need to define additional site-specific conditions and materials parameters. Thus, both the European and the LIFE-365 approaches are more simplified, based largely on the diffusion mechanism, with some modifications for convection effects.

The choice and guidelines for determining the various input parameters and variables in the models are outlined below for the apparent diffusion coefficient, chloride concentration at the concrete surface, and the critical chloride concentration required for depassivation. The information given is for European models developed within the context of EU research projects and the *fib* Model Code, and in North America for the LIFE-365 model developed within the context of the activities of ACI. LIFE-365 comes in the form of software, which allows one to introduce values of variables that are different from the recommended ones. This is obviously also true for the European model. The Japanese specification goes further to provide more specific numerical information.

1. *Apparent diffusion coefficient*: The apparent diffusion coefficient of chlorides at time t can be described in terms of the coefficient in standard conditions, with modifications taking into account the temperature of the surrounding environment and the effect of aging (continued hydration reduces the coefficient), as described in the *fib* Model Code for Service Life Design (2006):

$$D_{app,C}(t) = k_e \cdot D_{RCM,0} \cdot A(t)$$ (5.34)

where $D_{RCM,0}$ is the diffusion coefficient in standard conditions, k_e is the environmental (temperature) effect, and $A(t)$ is the aging (time) effect.

The Japanese Standard Specifications for Concrete Structures—2007: Design propose to consider also the effect of cracking on the diffusion coefficient, taking into account the crack width as well as crack spacing (see later for the effect of cracking).

Each component in these equations has been quantified by values proposed in guides or commentaries to codes and specifications.

a. Diffusion coefficient in standard conditions, $D_{RCM,0}$

i. *fib Model Code*: The EU recommendations are based on determinations by means of accelerated migration tests. Guiding values are provided in Table 5.11, with a s.d. of 20%. However, it should be noted that these are really average values, since there is considerable variability within each "cement type."

ii. *Japan*: The estimated diffusion coefficient, D_p, is suggested to be a function of w/c ratio.

For normal Portland cement:

$$\log D_p = -3.9\left(\frac{w}{c}\right)^2 + 7.2\left(\frac{w}{c}\right) - 2.5; \text{ (cm}^2\text{/year)}$$ (5.35)

Table 5.11 Recommended values for diffusion coefficients

| Cement type | Diffusion coefficient (m^2/s) at equivalent w/c ratios of | | | | | |
	0.35	0.40	0.45	0.50	0.55	0.60
CEM I 42.5R	–	8.9×10^{-12}	10.0×10^{-12}	15.8×10^{-12}	19.7×10^{-12}	25.0×10^{-12}
CEM I 42.5R + FA (k = 0.5)	–	5.6×10^{-12}	6.9×10^{-12}	9.0×10^{-12}	10.9×10^{-12}	14.9×10^{-12}
CEM I 42.5R + SF (k = 2.0)	4.4×10^{-12}	4.8×10^{-12}	–	–	5.3×10^{-12}	–
CEM III/B 42.5	–	1.4×10^{-12}	1.9×10^{-12}	2.8×10^{-12}	3.0×10^{-12}	3.4×10^{-12}

Source: Adapted from *fib* Model Code for Service Life Design (*fib* MC-SLD), 2006, *fib* Bulletin no. 34, 2006, *fib* secretariat, Case postale 88, CH-1015, Lausanne, Switzerland.

For blast-furnace slag cement:

$$\log D_p = -3.0\left(\frac{w}{c}\right)^2 + 5.4\left(\frac{w}{c}\right) - 2.2; \text{(cm}^2/\text{year)} \qquad (5.36)$$

The characteristic values taken for design, D_k, include also a partial factor, γ_p:

$$D_p \cdot \gamma_p \leq D_k \qquad (5.37)$$

where γ_p is a factor taking into account the inaccuracy in determining D_p, lying between 1.0 and 1.3.

  iii. *LIFE-365*: LIFE-365 provides equations accounting for the influence of *w/cm* ratio and silica fume content:

    w/cm ratio:

$$D_{28} = 1 \times 10(-12.06 + 2.40\,w/cm), \text{(m}^2/\text{s)} \qquad (5.38)$$

Silica fume content:

$$D_{SF} = D_{PC} \cdot e^{-0.165 \cdot SF} \qquad (5.39)$$

Graphical descriptions of these equations are presented in Figures 5.11 and 5.12.

  b. *Effect of cracking*: The Japanese specifications consider the influence of cracking on the characteristic diffusion coefficient, D_k, to determine the diffusion coefficient for design, D_d, in the presence of a crack having a width w:

$$D_d = \gamma_c \cdot D_k + \left(\frac{w}{l}\right)\left(\frac{w}{w_a}\right)^2 \cdot D_o \qquad (5.40)$$

where γ_c is the material factor for concrete, taken usually as 1.0, but 1.3 for the top face of the structure (if the concrete quality in the structure is the same as that of specimens cured under standard conditions, the material factor for all parts may be taken as 1.0); D_k is the characteristic diffusion coefficient (cm²/year); D_o is a constant representing the effect of cracks on transport of chlorides and is considered only in cases when flexural cracks are allowed (in general, it may be taken to be 200 cm²/year); w is the calculated flexural crack width; w_a is the limit value of crack width (mm) having values proportional to the concrete

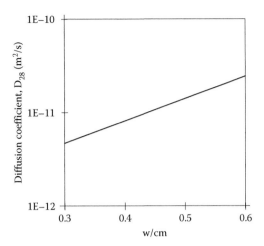

Figure 5.11 Effect of w/cm ratio on the base diffusion coefficient. (Adapted from LIFE-365, 2012, *Service Life Prediction Model and Computer Program for Predicting the Service Life and Life-Cycle Cost of Reinforced Concrete Exposed to Chlorides Version 2.1*, January 7, Produced by the LIFE-365™ Consortium II, The Silica Fume Association www.life-365.org and www.silicafume.org, reprinted with permission.)

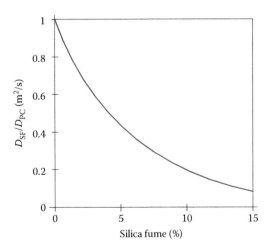

Figure 5.12 Effect of silica fume on the base diffusion coefficient. (Adapted from LIFE-365, 2012, *Service Life Prediction Model and Computer Program for Predicting the Service Life and Life-Cycle Cost of Reinforced Concrete Exposed to Chlorides Version 2.1*, January 7, Produced by the LIFE-365™ Consortium II, The Silica Fume Association www.life-365.org and www.silicafume.org, reprinted with permission.)

cover of 0.5%, 0.4%, and 0.3% of the concrete cover for normal, corrosive, and severely corrosive environmental conditions, respectively; and w/l is the ratio between crack width and crack spacing, which can be calculated following structural design guidelines.

Design diffusion coefficients (in cm²/year) calculated according to the Japanese specifications as a function of w/c ratio and crack width are presented in Table 5.12 and Figure 5.13.

Table 5.12 Design diffusion coefficient D_d (in cm²/year) calculated from crack width and water–cement ratio according to Japanese standard specifications for concrete structures

w/c ratio	Without crack	Crack width w (mm)		
		0.10	0.15	0.20
0.30	0.20	0.271	0.429	0.751
0.35	0.35	0.413	0.567	0.873
0.40	0.57	0.653	0.782	1.08
0.45	0.89	0.95	1.10	1.38
0.50	1.33	1.39	1.53	1.81
0.55	1.90	1.96	2.10	2.37
0.60	2.60	2.65	2.79	3.05
0.65	3.39	3.45	3.58	3.84
0.70	4.24	4.29	4.42	4.68

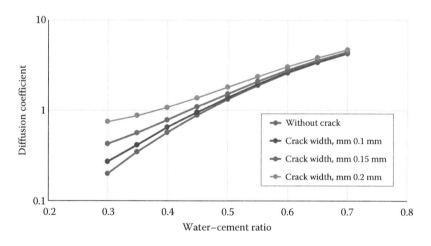

Figure 5.13 Design values for diffusion coefficient (in cm²/year units) as a function of w/c ratio and crack width calculated according to the Japanese Standard Specifications for Concrete Structures.

Assumptions:

Ordinary Portland cement

Allowable crack width 0.20 mm

Material coefficient for concrete, γ_c, 1.0

Constant indicating the effect of cracking on the transfer of chloride ions in concrete, D_o, is 200 cm²/year.

c. *Environmental (temperature) effect, k_e:* The effect of the environment on the diffusion coefficient is largely due to temperature and has been similarly described in the EU and LIFE-365 models:

   i. *fib Model Code:*

$$K_e = \exp\left\{ b_e \left[\frac{1}{T_{ref}} - \frac{1}{T_{real}} \right] \right\} \tag{5.41}$$

with b_e having a value of 4800 and s.d. of 700, and T_{ref} is the standard test temperature (°K) and T_{real} is the temperature of the structural element or the ambient air (°K).

   ii. *LIFE-365:*

$$K_e = \exp\left[\frac{U}{R}\left(\frac{1}{T_{ref}} - \frac{1}{T_{real}} \right) \right] \tag{5.42}$$

where U is the activation energy of the diffusion process (35,000 J/mol), R is a universal gas constant, and T is absolute temperature.

The temperature should be chosen based upon local weather information such as shown in Figure 5.14.

d. Change in diffusion coefficient due to aging, $A(t)$

   i. *fib Model Code:* The diffusion coefficient reduces over time due to continued hydration.

   This has been described in the EU in terms of Equation 5.43:

$$A(t) = \left(\frac{t_o}{t} \right)^a \tag{5.43}$$

where a is the aging exponent (diffusion decay index) and t_o is the reference point of time.

Guidelines for statistical quantification of the exponent are given in Table 5.13, for a reference time of 28 days.

   ii. *LIFE-365:* A similar equation is used in LIFE-365 with the following recommended values (Table 5.14) for the aging exponent a, which is highly sensitive to the composition of the binder.

$$a = 0.2 + 0.4\left(\frac{\%FA}{50} + \frac{\%SG}{70} \right) \tag{5.44}$$

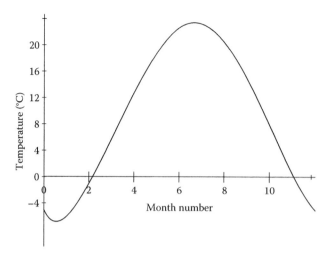

Figure 5.14 Example of annual temperature profile to be used in Equation 5.41 for the diffusion coefficient. (Adapted from LIFE-365, 2012, *Service Life Prediction Model and Computer Program for Predicting the Service Life and Life-Cycle Cost of Reinforced Concrete Exposed to Chlorides Version 2.1*, January 7, Produced by the LIFE-365™ Consortium II, The Silica Fume Association www.life-365.org and www.silicafume.org, reprinted with permission.)

The higher aging exponent for the mineral admixture expresses their continued pozzolanic activity which refines the pore space and improved chloride binding capacity, which leads to a marked reduction in the diffusion coefficient over time, as demonstrated in Figure 5.15. Note that the diffusion coefficient is assumed to be constant after 25 years.

Table 5.13 Guidelines for statistical parameters to Equation 5.42

Concrete	Aging exponent, a (parameters for β-distribution)[e]
Portland cement concrete, CEM I 0.40 < w/c < 0.60	$m^a = 0.30; s^b = 0.12; a^c = 0.0; b^d = 1.0$
Portland fly ash cement concrete, $f > 0.20c$; $k = 0.50; 0.40 < w/c_{eq} < 0.60$	$m^a = 0.60; s^b = 0.15; a^c = 0.0; b^d = 1.0$
Blast-furnace slag cement concrete, CEM III/B; 0.40 < w/c < 0.60	$m^a = 0.45; s^b = 0.20; a^c = 0.0; b^d = 1.0$

Source: Adapted from *fib* Model Code for Service Life Design (*fib* MC-SLD), 2006, *fib* Bulletin no. 34, 2006, *fib* secretariat, Case postale 88, CH-1015, Lausanne, Switzerland.

[a]   *m*—mean value.
[b]   *s*—standard deviation.
[c]   *a*—lower bound.
[d]   *b*—upper bound.
[e]   Quantification can be applied for the exposure classes: splash zone, tidal zone, and submerged zone.

Table 5.14 Recommended values for design in the LIFE-365 model

Binder	a	$D_{28d} \times 10^{-13} m^2/s$	$D_{10y} \times 10^{-13} m^2/s$	$D_{25y} \times 10^{-13} m^2/s$
PC	0.20	79	30	25
30% slag	0.37	79	13	9.3
40% fly ash	0.52	79	6.3	3.9

Source: Adapted from LIFE-365, 2012, *Service Life Prediction Model and Computer Program for Predicting the Service Life and Life-Cycle Cost of Reinforced Concrete Exposed to Chlorides, Version 2.1*, January 7, Produced by the LIFE-365™ Consortium II, The Silica Fume Association www.life-365.org and www.silica-fume.org, reprinted with permission.

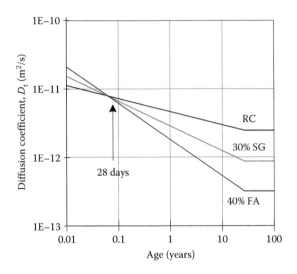

Figure 5.15 Effect of time on reduction in diffusion coefficient. (Adapted from LIFE-365, 2012, *Service Life Prediction Model and Computer Program for Predicting the Service Life and Life-Cycle Cost of Reinforced Concrete Exposed to Chlorides, Version 2.1*, January 7, Produced by the LIFE-365™ Consortium II, The Silica Fume Association www.life-365.org and www.silicafume.org, reprinted with permission.)

2. Critical chloride content
   a. *fib Model Code*: The lower boundary of the critical chloride content, C_{crit}, is usually taken as 0.20% by wt. of cement. Taking this as a lower bound, a critical chloride distribution function that provides sufficiently good agreement with test results (Maage et al. 1996) was suggested, being a β-distribution with 0.20% lower bound, and a mean of 0.60%.

   Thus, the critical chloride content can be quantified as follows: β-distributed, mean of 0.6, s.d. of 0.15, lower bound of 0.2, and higher bound of 2.

b. *Japan*: The Japanese specifications propose a reference to measurement results or test results of similar structures. If these are not available, it is recommended to use the value of 1.2 kg/m³. For structures exposed to freezing and thawing, it is suggested to consider lower values.

c. *LIFE-365*: The critical chloride content is taken as 0.05% by weight of concrete, and its value is adjusted when special protective means are used, such as use of a corrosion inhibitor in which the action is quantified by an increase in the critical chloride value (Table 5.15).

When 316 stainless steel reinforcing is used, LIFE-365 recommends the use of a critical chloride value of 0.5%, which is 10 times higher than that for conventional steel.

The guidelines for epoxy-coated steel are quite different. It is assumed that there is no influence on the critical chloride value but rather an increase of 6–20 years in the propagation period until damage occurs.

3. Chloride surface content buildup

a. *fib Model Code*: The chloride content on the surface depends on the environmental conditions and the properties of the concrete. Various tests and data are available in the literature to account for this value. Examples are given in the *fib* Model Code, shown in Figure 5.16 for Portland cement concrete (*w/c* of 0.50 and cement content of 300 kg/m³).

Suggested values for buildup in various environmental conditions are provided in Table 5.16, in terms of the maximum attainable level and its buildup rate.

Table 5.15 Chloride threshold values recommended in LIFE-365 (2012) for calcium nitrite inhibitor

CNI[a] dose (L/m³)	Threshold chloride (% wt. of concrete)
0	0.05
10	0.15
15	0.24
20	0.32
25	0.37
30	0.40

Source: Adapted from LIFE-365, 2012, *Service Life Prediction Model and Computer Program for Predicting the Service Life and Life-Cycle Cost of Reinforced Concrete Exposed to Chlorides Version 2.1*, January 7, Produced by the LIFE-365™ Consortium II, The Silica Fume Association www.life-365.org and www.silicafume.org, reprinted with permission.

a   Calcium nitrite inhibitor.

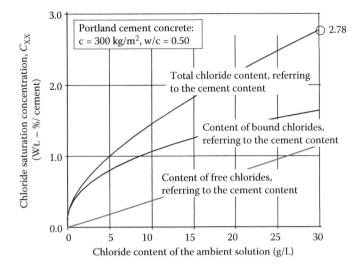

Figure 5.16 Surface chloride concentration as a function of the ambient solution. (Adapted from *fib* Model Code for Service Life Design (*fib* MC-SLD), 2006, *fib* Bulletin no. 34, 2006, *fib* secretariat, Case postale 88, CH-1015, Lausanne, Switzerland.)

b. *Japan*: The Japanese specifications recommend the use of values based on the performance data of similar structures that accurately express the environmental conditions. If these are not available, the chloride surface concentration may be determined from Table 5.17, which corresponds to a height of 1 m and a distance of 25 m from the shoreline.

c. *LIFE-365*: LIFE-365 provides guidelines for the buildup of chloride on the concrete surface, assuming that there is a maximum value that can be achieved under specific conditions and that there is a gradual buildup toward this value. This is shown in Figure 5.17 for the reference concrete (PC) as well as for concretes in which the surface has been treated by protective measures such as membranes and sealers which slow down this buildup.

Table 5.16 Buildup rates and maximum surface concentrations at various zones, based on data in the model code

	Buildup rate (%/year)	Maximum (% wt. of concrete)
Marine splash zone	Instantaneous	0.8
Marine spray zone	0.10	1.0
Within 800 m of the sea shore	0.04	0.6

Table 5.17 Chloride ion concentration C_0 (kg/m³) at the concrete surface

Regions		Splash zone	Distance from coast (km)				
			Near shoreline	0.1	0.25	0.1	1.0
Regions with high airborne chloride concentration	Hokkaido, Tohoku, Hokuriku, Okinawa	13.0	9.0	4.5	3.0	2.0	1.0
Regions with low airborne chloride concentration	Kanto, Kinki, Chugoku, Shikoku, Kyushu		4.5	2.5	2.9	1.5	1.0

Source: Japanese Standard Specifications for Concrete Structures, 2007, *Design, JSCE Guidelines for Concrete No. 15,* Japan Society of Civil Engineers.

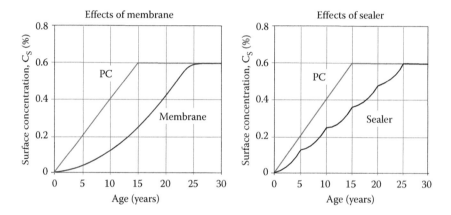

Figure 5.17 Effect of various surface treatments on the buildup of chlorides, % by mass of concrete. (Adapted from LIFE-365, 2012, *Service Life Prediction Model and Computer Program for Predicting the Service Life and Life-Cycle Cost of Reinforced Concrete Exposed to Chlorides Version 2.1,* January 7, Produced by the LIFE-365™ Consortium II, The Silica Fume Association www.life-365.org and www.silicafume.org, reprinted with permission.)

5.5.1.3 Residual service life-cycle estimation

Relations and data of the kind reviewed here enable the determination of service life for a given depth of cover or for the design to achieve a specified service life, by controlling the concrete quality and depth of cover. For existing structures, the relations outlined in this chapter can provide the basis for calculating the residual service life.

For design of new buildings, there is a need to take into account statistical considerations, by either the full probabilistic approach or the partial factor approach.

The Japanese specifications provide guidelines for partial factor design by setting the limit state requirement as follows:

$$\gamma_i \left(\frac{C_d}{C_{lim}} \right) < 1.0 \tag{5.45}$$

where γ_i is a structure factor which can be taken as 1.0 for ordinary structures and 1.1 for important structure; C_{lim} is the threshold chloride concentration that should be determined from data on similar structures or, if not available, 1.2 kg/m³; and C_d is the design value of chloride concentration estimated from the diffusion equation:

$$C_d = \gamma_{cl} \cdot C_o \left\{ 1 - \mathrm{erf} \left(\frac{0.1 \cdot c}{2(D_d \cdot t)^{1/2}} \right) \right\} \tag{5.46}$$

where C_o is the concentration of chloride ions on the concrete surface (kg/m³); γ_{cl} is a factor taking into account the scatter in Cd (it may be taken as 1.3); c is the concrete cover (mm); c_d is the design value of concrete cover used for durability verification (mm), $c_d = c - \Delta_{ce}$, Δ_{ce}—construction inaccuracy (mm); t is the service life design (years); and D_d is the design value of chloride diffusion coefficient (cm²/year).

This approach and the accompanying recommended data can be used to design for a specified service life. Examples of combinations that meet the requirements are presented in Table 5.18, outlining the maximum diffusion coefficient required to provide the specified design service life.

A graphical representation showing the combination of design diffusion coefficient and design concrete cover based on the above equations is provided in Figure 5.18, taken from the guidelines of the Japanese Standard Specifications for Concrete Structures.

5.5.2 Frost-induced internal damage

The limit state equation for frost resistance formulated in the *fib* Model Code for Service Life Design (2006) is based on the concept that the degree of saturation at any time t, $S_{ACT}(t)$, is smaller than the critical degree of saturation, S_{CR}, during the entire service life:

$$f(S_{CR}, S_{ACT}(t)) = S_{CR} - S_{ACT}(t) > 0 \tag{5.47}$$

The limit state equation is based on the absorption of water into the air-void system to reach levels exceeding the critical degree of saturation, which is considered to be a material property.

The critical degrees of saturation can be determined in lab tests where the concrete is conditioned at different degrees of saturation and then

Table 5.18 Maximum design chloride diffusion coefficient, D_d, for achieving specified service life in various exposure conditions

Life time (years)	Design concrete cover (mm)									
	25	30	35	40	50	60	70	100	150	200
Splash zone, $C_0 = 13$ kg/m³										
20	–	–	–	0.123	0.192	0.276	0.376	0.767	1.720	3.070
30	–	–	–	–	0.128	0.184	0.250	0.511	1.150	2.040
50	–	–	–	–	–	0.110	0.150	0.307	0.690	1.230
100	–	–	–	–	–	–	–	0.153	0.345	0.613
Near shore line, $C_0 = 9$ kg/m³										
20	–	–	0.115	0.150	0.295	0.338	0.460	0.989	2.110	3.750
30	–	–	–	0.100	0.156	0.225	0.307	0.626	1.410	2.500
50	–	–	–	–	–	0.135	0.184	0.375	0.845	1.500
100	–	–	–	–	–	–	–	0.188	0.422	0.751
0.1 km from coast, $C_0 = 4.5$ kg/m³										
20	–	0.140	0.191	0.249	0.389	0.561	0.763	1.560	3.500	6.230
30	–	–	0.127	0.166	0.260	0.374	0.509	1.040	2.340	4.150
50	–	–	–	–	0.156	0.224	0.305	0.623	1.400	2.490
100	–	–	–	–	–	0.112	0.153	0.311	0.700	1.250
0.25 km from coast, $C_0 = 3$ kg/m³										
20	0.150	0.216	0.295	0.385	0.601	0.866	1.180	2.400	5.410	9.620
30	0.100	0.144	0.196	0.256	0.401	0.572	0.785	1.600	3.610	6.410
50	–	–	0.118	0.154	0.240	0.346	0.471	0.962	2.160	3.850
100	–	–	–	–	0.120	0.173	0.236	0.481	1.080	1.920
0.5 km from coast, $C_0 = 2$ kg/m³										
20	0.288	0.414	0.564	0.737	1.150	1.660	2.226	4.610	10.400	18.400
30	0.192	0.276	0.376	0.491	0.768	1.110	1.500	3.070	6.910	12.300
50	0.115	0.166	0.226	0.295	0.461	0.663	0.903	1.840	4.140	7.370
100	–	–	0.113	0.147	0.230	0.332	0.451	0.920	2.070	3.630
1 km from coast, $C_0 = 1.5$ kg/m³										
20	0.620	0.803	1.220	1.590	2.450	3.570	4.860	9.920	22.300	39.700
30	0.413	0.595	0.810	1.060	1.650	2.380	3.240	6.610	14.900	26.400
50	0.248	0.357	0.486	0.635	0.992	1.430	1.940	3.970	8.930	15.900
100	0.124	0.179	0.243	0.317	0.406	0.714	0.972	1.980	4.460	7.930

Source: Japanese Standard Specifications for Concrete Structures, 2007, *Design, JSCE Guidelines for Concrete No. 15*, Japan Society of Civil Engineers.

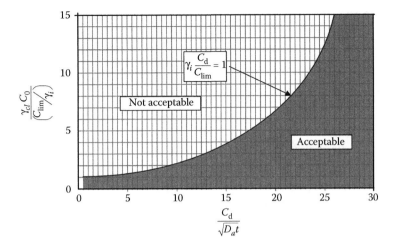

Figure 5.18 Graphical presentation of combination of design diffusion coefficient and design concrete cover based on Japanese Standards Specification for Concrete Structures.

exposed to freeze–thaw cycles. The dynamic modulus of elasticity is measured before and after the cycling, and the lowest degree of saturation at which a marked decline in the modulus is observed is defined as the critical degree of saturation.

The water absorption into the concrete can be described in terms of Equation 5.48:

$$S_{ACT}(t) = S_n + e \cdot t_{eq}^d \tag{5.48}$$

where t_{eq} is the equivalent time of wetness; and S_n, and e and d are materials parameter and exponents, respectively.

The design by the full probabilistic approach or the partial factor approach is based on the limit state equation (Equation 5.48), adopting the relations for the depth of penetration such as Equation 5.10 for the full probabilistic approach and Equations 5.11 through 5.13 for the partial factor approach.

The Japanese specifications suggest a different limit state, which is based on the measurement of the dynamic modulus of elasticity in accelerated freeze–thaw tests. The design relative dynamic modulus of elasticity, E_d, should exceed $\gamma_i \cdot E_{min}$, where γ_i is a structure factor and E_{min} is the minimum limit value of relative dynamic modulus of elasticity in freezing and thawing, required to meet freeze–thaw resistance performance in a concrete structure:

$$\gamma_i \left(\frac{E_{min}}{E_d} \right) \leq 1.0 \tag{5.49}$$

Table 5.19 Minimum limit value of relative dynamic modulus of elasticity to meet freeze–thaw resistance performance requirements for concrete structure

Climate		Severe weather conditions or frequent freezing–thawing action		Not so severe weather conditions, atmospheric temperature rarely drop to below 0°C	
Section		Thin (<20 cm)	General	Thin (<20 cm)	General
Exposure	(1) Immersed in water or often saturated with water	85	70	85	60
	(2) Not covered in (1) and subjected to normal exposure conditions	70	60	70	60

Source: Japanese Standard Specifications for Concrete Structures, 2007, *Design, JSCE Guidelines for Concrete No. 15*, Japan Society of Civil Engineers.

where γ_i is a factor representing the importance of the structure, taken usually as 1.0 and may be increased to 1.1 for important structures; E_d is the design value of relative dynamic modulus of elasticity in freeze–thaw test ($=E_k/\gamma_c$), where E_k is the characteristic value of relative dynamic modulus of elasticity in freeze–thaw test, conducted according to the Japanese standard JIS A 1148; and E_{min} is the critical minimum value of relative dynamic modulus of elasticity to ensure satisfactory performance of the structure in freezing–thawing action. Recommended values are provided in Table 5.19.

5.5.3 Chemical attack

The Japanese specifications suggest design for chemical attack, which is based on the concept of chemical penetration depth, applying relations of the nature of Equations 5.11 through 5.13. However, it was stated that at the time of the publication of the specifications there was not sufficient knowledge for accepted quantitative models. This is also the view expressed in the *fib* Model Code 2010.

5.6 DEEMED-TO-SATISFY APPROACH

The deemed-to-satisfy approach is prescriptive in nature and as such it is very much related to national standards reflecting local conditions and experience. A prescriptive approach that has more generalized characteristics can be found in the Eurocode 2 (EN 1992) in which the concepts of EN 206 (EN 206, 2013) were adopted to define environmental exposure classes in terms of the environmental processes which may be invoked (Tables 5.20 and 5.21).

Table 5.20 Exposure classes related to environmental conditions in accordance with EN 206

Class designation	Description of the environment	Informative examples where exposure classes may occur
No risk of corrosion or attack		
X0	For concrete without reinforcement or embedded metal: all exposures except where there is freeze–thaw, abrasion or chemical attack For concrete with reinforcement or embedded metal: very dry	Concrete inside buildings with very low air humidity
Corrosion induced by carbonation		
XC1	Dry or permanently wet	Concrete inside buildings with low air humidity Concrete permanently submerged in water
XC2	Wet, rarely dry	Concrete surfaces subject to long-term water contact Many foundations
XC3	Moderate humidity	Concrete inside buildings with moderate or high air humidity External concrete sheltered from rain
XC4	Cyclic wet and dry	Concrete surfaces subject to water contact, not within exposure class XC2
Corrosion induced by chlorides		
XD1	Moderate humidity	Concrete surfaces exposed to airborne chlorides
XD2	Wet, rarely dry	Swimming pools Concrete components exposed to industrial waters containing chlorides
XD3	Cyclic wet and dry	Parts of bridges exposed to spray containing chlorides Pavements Car park slabs
Corrosion induced by chlorides from sea water		
XS1	Exposed to airborne salt but not in direct contact with sea water	Structures near to or on the coast
XS2	Permanently submerged	Parts of marine structures

(Continued)

Table 5.20 (Continued) Exposure classes related to environmental conditions in accordance with EN 206

Class designation	Description of the environment	Informative examples where exposure classes may occur
XS3	Tidal, splash, and spray zones	Parts of marine structures
Freeze–thaw attack		
XF1	Moderate water saturation, without deicing agent	Vertical concrete surfaces exposed to rain and freezing
XF2	Moderate water saturation, with deicing agent	Vertical concrete surfaces of road structures exposed to freezing and airborne deicing agents
XF3	High water saturation, without deicing agents	Horizontal concrete surfaces exposed to rain and freezing
XF4	High water saturation, with deicing agents or sea water	Road and bridge decks exposed to deicing agents Concrete surfaces exposed to direct spray containing deicing agents and freezing Splash zone of marine structures exposed to freezing
Chemical attack		
XA1	Slightly aggressive chemical environment according to EN 206-1, Table 2	Natural soils and groundwater
XA2	Moderately aggressive chemical environment according to EN 206-1, Table 2	Natural soils and groundwater
XA3	Highly aggressive chemical environment according to EN 206-1, Table 2	Natural soils and groundwater

Each of the exposure classes in Table 5.20 is subdivided, according to the severity of the degradation process in each class, depending mainly on the moisture conditions. A rough indication of the effect of moisture conditions on the intensity of the degradation process is presented in Table 5.22 (Table 5.8-1 in *fib*, 1999). In very dry concrete, the degradation reactions are very slow and they are most intensive when the relative humidity in the concrete is in the range of 75%–98%.

The requirement of the concrete for each of the exposure classes is outlined in Appendix F to EN 206, Table F1 (EN 206, 2013), which is a nonbinding recommendation to be considered for adoption in a national annex (Table 5.23). The values in this table are based on the assumption of an intended design working life of 50 years. The minimum strength classes were derived from the relationship between water–cement ratio and the strength class of concrete made with cement of strength class 32.5.

Table 5.21 Limiting values for exposure classes for chemical attack from natural soil and groundwater in accordance with EN 206

The aggressive chemical environments classified below are based on natural soil and groundwater at water/soil temperatures between 5°C and 25°C and a water velocity sufficiently slow to approximate to static conditions.

The most onerous value for any single chemical characteristic determines the class.

Where two or more aggressive characteristics lead to the same class, the environment shall be classified into the next higher class, unless a special study for this specific case proves that it is not necessary.

Chemical characteristic	Reference test method	XA1	XA2	XA3
Groundwater				
SO_4^2 (mg/L)	EN 196-2	>200 and <600	>600 and <3000	>3000 and <6000
pH	ISO 4316	<6.5 and >5.5	<5.5 and >4.5	<4.5 and >4.0
CO_2 (mg/L) aggressive	prEN 13577:1999	>15 and <40	>40 and <100	>100 up to saturation
NH_4^+ (mg/L)	ISO 7150-1 or ISO 7150-2	>15 and <30	>30 and <60	>60 and <100
Mg^{2+} (mg/L)	ISO 7980	>300 and <1000	>1000 and <3000	>3000 up to saturation
Soil				
SO_4^2 (mg/kg[a]) total	EN 196-2[b]	>2000 and <3000[c]	>3000[c] and <12,000	>12,000 and <24,000
Acidity (mL/kg)	DIN 4030-2	>200 Baumann–Gully	*Not encountered in practice*	

[a] Clay soils with a permeability below 10 m[5]/s may be moved into a lower class.
[b] The test method prescribes the extraction of SO_4^2 by hydrochloric acid; alternatively, water extraction may be used, if experience is available in the place of use of the concrete.
[c] The 3000 mg/kg limit shall be reduced to 2000 mg/kg, where there is a risk of accumulation of sulfate ions in the concrete due to drying and wetting cycles or capillary suction.

Table 5.22 Estimation of the influence of moisture on corrosion and degradation processes

Corrosion/degradation mechanism	Durability performance risk at moisture conditions, quantified in terms of relative humidity (%)				
	<45%	45%–65%	65%–85%	85%–98%	Saturated
Carbonation	Low	High	Medium	Low	None
Steel corrosion	Low	Medium	High	Medium	Low/none
Frost	Low/none	Low/none	Low/none	Medium	High
Chemical attack	Low/none	Low/none	Low/none	Medium	High

Source: Based on information in CEB 1992, *Durability of Concrete Structures—Design Guide*, Thomas Telford Services, Ltd., London, UK, 1992, quoted from *fib*, 1999, *Textbook on Behavior, Design and Performance—Updated Knowledge of the CEB/FIP Model Code 1990*, Vol. 3, Bulletin 3.

Table 5.23 Recommended limiting values for composition and properties of concrete in EN 206

| | No risk of corrosion or attack | Carbonation-induced corrosion | | | | Chloride-induced corrosion | | | | | | Freeze–thaw attack | | | | Aggressive chemical environments | | |
| | | | | | | Sea water | | | Chloride other than from sea water | | | | | | | | | |
	X0	XC1	XC2	XC3	XC4	XS1	XS2	XS3	XD1	XD2	XD3	XF1	XF2	XF3	XF4	XA1	XA2	XA3
Maximum w/c	–	0.65	0.60	0.55	0.50	0.50	0.45	0.45	0.55	0.55	0.45	0.55	0.55	0.50	0.45	0.55	0.50	0.45
Minimum strength class	C12/15	C20/25	C25/30	C30/37	C30/37	C30/37	C35/45	C35/45	C30/37	C30/37	C35/45	C30/37	C25/30	C30/37	C30/37	C30/37	C30/37	C35/45
Minimum cement content (kg/m³)	–	260	280	280	300	300	320	340	300	300	320	300	300	320	340	300	320	360
Minimum air content (%)	–	–	–	–	–	–	–	–	–	–	–	–	4.0[a]	4.0[a]	4.0[a]	–	–	–
Other requirements												Aggregate in accordance with EN 12620 with sufficient freeze–thaw resistance				Sulfate-resisting cement[b]		

a Where the concrete is not air entrained, the performance of concrete should be tested according to an appropriate test method in comparison with a concrete for which freeze–thaw resistance for the relevant exposure class is proven.

b When SO_4^{2-} leads to exposure Classes XA2 and XA3, it is essential to use sulfate-resisting cement. Where cement is classified with respect to sulfate resistance, moderate or high sulfate-resisting cement should be used in exposure Class XA1 (and in exposure Class XA2 when applicable) and high sulfate-resisting cement should be used in exposure Class XA3.

Table 5.24 Values of minimum cover thickness, $c_{min,dur}$, according to Eurocode 2

	Environmental Requirement for $c_{min,dur}$ (mm)						
	Exposure cass according to Table 5.19						
Structural class	X0	XCI	XC2/XC3	XC4	XDI/XSI	XD2/XS2	XD3/XS3
S1	10	10	10	15	20	25	30
S2	10	10	15	20	25	30	35
S3	10	10	20	25	30	35	40
S4	10	15	25	30	35	40	45
S5	15	20	30	35	40	45	50
S6	20	25	35	40	45	50	55

Source: Adapted from EN, 1992, Design of Concrete Structures, Table 4.4N.

Thus, the limiting values for the maximum water–cement ratio and the minimum cement content apply in all cases, while the requirements for the strength class may be additionally specified.

The prescriptive type of standard also specifies the cover thickness. For the strength classes in Table 5.23, the cover thickness required for the different exposure classes to achieve 50 years of service life is presented in Table 5.24, where structural class 4 is related to the strength class in Table 5.23.

Eurocode 2 specifies the conditions for achieving 100 years of service life, by increasing the concrete strength class by two levels, as well as providing relaxed requirement for strength class (reduction by one strength level) if the production processes are especially well controlled, providing either accurate reinforcement position (e.g., precast concrete) or special quality control of the concrete production.

The nominal concrete cover thickness, which is the value specified in the drawings is defined as the minimum cover thickness, c_{min}, plus an allowance in design for deviation, Δc_{dev}. The minimum cover value c_{min} is provided to ensure safe transmission of bond forces ($c_{min,b}$), protection of the steel against corrosion (durability, $c_{min,dur}$), and adequate fire resistance. The greater value for c_{min} satisfying the requirement for both bond and environmental conditions shall be taken. The recommended value for Δc is 10 mm. It can be reduced if the quality assurance of the fabrication is high (in the range of 5–10 mm) and should be increased if the concrete is cast against uneven surface or the quality control is poor. If the nominal cover thickness is considered as the average value of a normal distribution, and taking Δc of 10 mm as the 5% fractile characteristic value (i.e., $\Delta c = 16,645$ s, where s is the s.d.), then the s.d. is about 6 mm. The actual s.d. on site can readily be determined by nondestructive tests.

EN 206 (2013), although a prescriptive standard, as seen in Table 5.23, provides another option, which might be considered as being a partial performance standard, as outlined in Section 5.2.5.3 of the standard, referring to the principles of the "Equivalent Concrete Performance Concept" (see also Section 5.1.4 in this chapter). It allows for changes in the requirement for minimum cement content and maximum water–cement ratio when addition is used or special cement is being employed. The concept is that the modified composition shall be proven to have equivalent performance to the reference concrete, with respect to its reaction to the environmental conditions.

5.7 LIFE-CYCLE COST ANALYSIS

The data used for calculating the service life can be further extended to calculate life-cycle cost based on assumptions with regards to costs of labor, cost of materials, cost of repair, and the service life of the repairs. This approach has been used to compare between different protection technologies in harsh weather, as demonstrated in Figure 5.19.

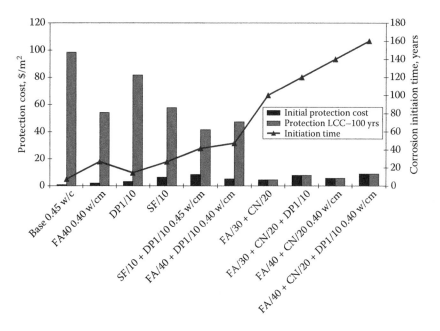

Figure 5.19 Prediction of costs (initial and life-cycle cost [LCC]) and corrosion initiation times for a marine wall with 100 years design life, using a single technology and integration of two and three technologies (FA, fly ash; SF, silica fume; DP, water repelling admixtures; CN, corrosion inhibitor). The average yearly temperature is 25°C, concrete cover is 50 mm, and w/cm for the base system is 0.45. (Adapted from Bentur, A., Berke, N. S., and Li, L., 2006, Integration of technologies for optimizing durability performance of reinforced concretes, in *Concrete Durability and Service Life Planning (ConcreteLife 06)*, K. Kovler (ed.), *Proceedings of the RILEM International Conference*, Israel, RILEM Publications PRO 46, Paris, France, pp. 247–258.)

REFERENCES

Andrade, C. and Tavares, F., 2012, *LIFEPRED Service Life Prediction Program.*

ACI 365.1R-00, 2000, Service Life Prediction, American Concrete Institute, Farmington Hills, MI.

ASTM C1202, 2012, *Standard Test Method for Electrical Indication of Concrete's Ability to Resist Chloride Ion Penetration*, ASTM International, West Conshohocken, PA.

Bentur, A., Berke, N. S., and Li, L., 2006, Integration of technologies for optimizing durability performance of reinforced concretes, in *Concrete Durability and Service Life Planning (ConcreteLife 06)*, K. Kovler (ed.), Proceeding RILEM International Conference, Israel, RILEM Publications PRO 46, Paris, France, pp. 247–258.

Boddy, A., Bentz, E., Thomas, M. D. A., and Hooton, R. D., 1999, An overview and sensitivity study of a multi-mechanistic chloride transport model, *Cement and Concrete Research*, 29, 827–837.

Canadian Standard CSA A23.1 and A23.2, 2004, *Concrete Materials and Methods of Concrete Construction*, Canadian Standards Association, Toronto, Canada.

DARTS, 2004a, Durable and Reliable Tunnel Structures: Deterioration Modelling, European Commission, Growths 2000, Contract G1RDCT-2000-00467, Project GrD1-25633.

DARTS, 2004b, Durable and Reliable Tunnel Structures: Data, European Commission, Growths 2000, Contract G1RD-CT-2000-00467, Project GrD1-25633.

de Schutter, G., 2009, How to evaluate equivalent concrete performance following EN 206-1: The Belgian approach, in *Concrte Life and Service Life Planning—ConcreteLife'09*, K. Kovler (ed.), RILEM Publications SARL, Bagneux, France, pp. 1–7.

DuraCrete, 2000, Probabilistic Performance Based Durability Design of Concrete Structures: Statistical Quantification of the Variables in the Limit State Functions, Report No. BE 95-1347.

EN, 1992, Design of Concrete Structures, Brussels: CEN (European Committee for Standardisation).

EN 206, 2013, Concrete—Specification, Performance, Production and Conformity, Brussels: CEN (European Committee for Standardisation), December.

fib, 1999, *Textbook on Behavior, Design and Performance—Updated Knowledge of the CEB/FIP Model Code 1990*, Vol. 3, Bulletin 3.

FIB Model Code 2010, 2013, International Federation for Structural Concrete (*fib*), Lausanne, Switzerland.

fib Model Code for Service Life Design (*fib* MC-SLD), 2006, *fib* Bulletin no. 34, 2006, *fib* secretariat, Case postale 88, CH-1015, Lausanne, Switzerland.

Folliard, K. J., Juenger, M., Schindler, A., Riding, K., Poole, J., Kallivokas, L. F., Slatnick, S., Whigham, J., and Meadows, J. L., 2008, *Prediction Model for Concrete Behavior—Final Report*, CTR-Technical Report 0-4653-1, Center for Transportation Research at the University of Texas, Austin, TX.

Gehlen, C. and Rahimi, A., 2011, *Compilation of Test Methods to Determine Durability of Concrete. A Critical Review*. RILEM Technical Committee TDC.

Helland, S., 2013, Design for service life: Implementation of FIB Model Code 2010 rules in the operational code ISO 16204, *Structural Concrete*, 14(1), 10–18.

ISO 2394, 1998, *General Principles on Reliability for Structures*, International Organization for Standards, Geneva, Switzerland.

ISO 2394, 2015, *General Principles on Reliability for Structures*, International Standards Organization, Geneva, Switzerland, 111pp.

ISO 13823, 2008, *General Principles on the Design of Structures for Durability*, ISO International Standard 13823, Geneva, Switzerland.

ISO 16204, 2012, *Durability—Service Life Design of Concrete Structures*, Geneva, Switzerland.

Japanese Standard Specifications for Concrete Structures, 2007, *Design, JSCE Guidelines for Concrete No. 15*, Japan Society of Civil Engineers.

Kasami, H., Izumi, I., Tomosawa, F., and Fukushi, I., 1986, Carbonation of concrete and corrosion of reinforcement in reinforced concrete, *Proceedings of First Australia-Japan Coordinated Workshop on Durability of Reinforced Concrete*, Building Research Institute, Japan, C4, October, pp. 1–13.

LIFE-365, 2012, *Service Life Prediction Model and Computer Program for Predicting the Service Life and Life-Cycle Cost of Reinforced Concrete Exposed to Chlorides Version 2.1*, January 7, Produced by the LIFE-365™ Consortium II.

LIFE-365® v2.2, 2013, available at: www.Life-365.org (free software).

LIFECON, 2003, Service life models instructions on methodology and application of models for the prediction of the residual service life for classified environmental loads and types of structures in Europe; LIFECON DELIVERABLE D 3.2, Sascha Lay, Prof. Dr.-Ing. Peter Schießl cbm—Technische niversität München, Shared-cost RTD project; Project acronym: LIFECON; Project full title: Life Cycle Management of Concrete Infrastructures for Improved Sustainability; Project duration: 01.01.2001 to 31.12.2003; Coordinator: Technical Research Centre of Finland (VTT), VTT Building Technology, Professor, Dr. Asko Sarja; Date of issue of the report: 30.11.2003; Project funded by the European Community.

Linger, L., Roziere, E., Loukili, A., Cussigh, F., and Rougeau, P., 2014, *Improving Performance of Concrete Structures*, Fourth International *fib* Congress, Mumbai, India.

Maage, M., Helland, S., Poulsen, E., Vennesland, Ø., and Carlsen, J. E., 1996, Service life prediction of existing concrete structures exposed to marine environment, *ACI Materials Journal*, 93(6), 100–100.

Martin-Peréz, B., Pantazopoulou, S. J., and Thomas, M. D. A., 1998, Finite element modelling of corrosion in highway structures, *Second International Conference on Concrete Under Severe Conditions—CONSEC '98*, Tromso, Norway, June.

Saetta, A., Scotta, R., and Vitaliani, R., 1993, Analysis of chloride diffusion into partially saturated concrete, *ACI Materials Journal*, 90(5), 441–451.

STADIUM®, available at: www.simcotechnologies.com (proprietary software).

Tang, L., 2008, Engineering expression of the ClinConc model for prediction of free and total chloride ingress in submerged marine concrete, *Cement and Concrete Research*, 38(8–9), 1092–1097.

Thompson, S. E., 1909, *Concrete in Railroad Construction*, The Atlas Portland Cement Company, New York, NY.

Yao, Y., Wang, Z., and Wang, L., 2012, Durability of concrete under combined mechanical load and environmental actions: A review, *Journal of Sustainable Cement-Based Materials*, 1(1–2), 2–15.

Chapter 6

Durability indicators (indexes) and their use in engineering practice

6.1 INTRODUCTION

Durability embraces a wide range of practical problems in concrete design and construction. A "durability problem" for a particular structure in one location may differ from that for a similar structure in another location, or for a different structure in the same location. For example, a structure in a marine environment with chlorides needs a different durability approach to one exposed to sulfate attack. Consequently, there is no "universal" solution to durability problems, and while environmental exposure classifications such as in EN 206 (2013) are helpful, they are inherently simplistic. Each situation must be evaluated on its own merits, and the designer has to exercise judgment in assessing the deterioration mechanisms operating on a structure in its locality over a given period of time. Therefore, designers also require good understanding of the dominant deterioration mechanisms in concrete structures as well as the influence of the exposure environments. These aspects are discussed in Chapter 4.

6.1.1 Durability problems in concrete structures; need for a new approach

The durability performance of construction materials has long been a concern for engineers. Durability is the ability of a material or structure to withstand the service conditions for which it is designed over a prolonged period without unacceptable deterioration. Concrete and reinforced concrete (RC) possess inherent chemical and dimensional stability in most environments, and RC structures are expected to be maintenance-free during their service life. However, these assumptions are in question given the weight of evidence of premature deterioration of concrete structures. Many modern concrete structures need substantial repairs and maintenance during their service life, with resultant costs to the economy reaching 3%–5% of GNP in some countries. Neville (1987) suggests that reasons for the widespread lack of durability include poor understanding of deterioration processes by designers, inadequate acceptance criteria of concrete on-site, and changes in cement properties and construction practices.

Deterioration of concrete begins almost immediately after casting as the hardened properties are influenced by early-age phenomena such as plastic cracking, bleeding, segregation, and thermal effects. In the hardened state, concrete may be affected by many internal and external factors that cause damage by physical and/or chemical mechanisms. Deterioration is often associated with ingress of external aggressive agents such that the near-surface concrete quality largely controls durability, especially for RC structures. The interaction between the various material and environmental elements influencing durability is shown in Figure 6.1.

Modern concrete construction is characterized by more consistent quality cement together with a greater variety of binder types and admixtures, higher allowable stresses, and faster concrete casting and setting times, with the consequence that concrete is more sensitive to abuse. This has contributed in part to the premature deterioration of modern concrete structures, which in turn has elicited more stringent specifications

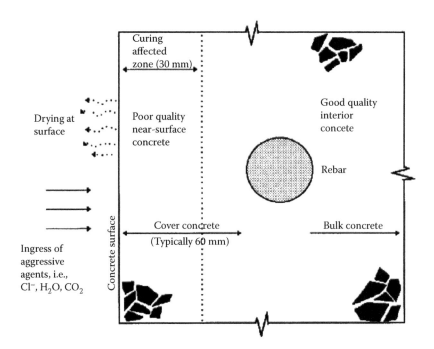

Figure 6.1 Schematic diagram of concrete protection of reinforcement. (Reproduced from Alexander, M. G., Mackechnie, J. R., and Ballim, Y., 2001, Use of durability indexes to achieve durable cover concrete in reinforced concrete structures, *Materials Science of Concrete*, J. P. Skalny and S. Mindess (eds.), American Ceramic Society, Vol. VI, pp. 483–511, with permission from Wiley Global.)

for concrete construction. These have unfortunately not always resulted in a corresponding improvement in durability performance due to a lack of understanding of what is required to ensure durability, as well as inadequate means of enforcing compliance with specifications during construction.

Most national codes and specifications are of the "recipe" type, setting limits on w/c ratios, cement content, cover, and so on, but without really addressing the issue of achieving adequate quality of the concrete. Further, it is difficult, if not impossible, to ensure on-site compliance with these specifications, since they generally comprise difficult-to-measure aspects of construction. (One notable exception is concrete cover to steel. Enforcing this simple expedient would probably cure 90% of current durability problems!)

In response to this situation, researchers are developing new tools to address durability problems. A powerful new set of tools involves the concept of "durability indicators" or "durability indexes." This chapter is largely given to a review of these new developments, philosophically and practically, with some general examples. However, first, some of the current and proposed approaches to addressing durability problems in concrete construction will be covered.

6.1.2 Traditional approach to concrete durability

Traditionally, durability has been addressed by regarding compressive strength as a "proxy" for durability. It is assumed that compressive strength reflects the durability properties on the basis that concrete with a suitable w/c ratio, properly proportioned and mixed, compacted, finished, and cured, will have acceptable compressive strength and should have sufficient durability. This approach for concrete durability is no longer adequate, considering advances in modern concrete technology such as the wide range of binders available, advances in structural design such as members of less bulk and mass, and the increasingly aggressive and polluted exposure environments. To make progress in improving concrete durability, we need to move away from the concept of strength-durability equivalence. Figure 6.2 shows the relationship between permeability (Darcy k value) and concrete compressive strength from measurements on actual as-built bridge structures in South Africa. There is no correlation, indicating that in actual construction, strength is not a suitable proxy for durability.

Another objection to the traditional approach is that fundamental factors that govern concrete properties, such as porosity, have different effects on durability and strength. Physical and chemical properties have significant influence on durability but a lesser effect on strength. For example, durability of RC structures is often governed not by the bulk strength, but by a surface penetrability property such as permeability. Further, curing has far

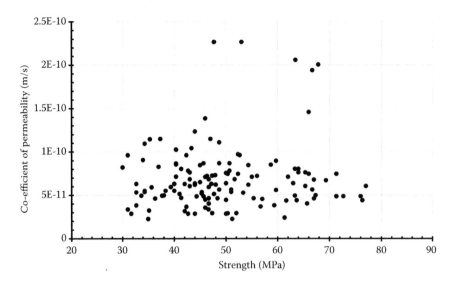

Figure 6.2 Comparison of coefficient of permeability values (*k*-value) with strength. (Reproduced from Nganga, G., Alexander, M. G., and Beushausen, H., 2013, Practical implementation of the durability index performance- based design approach, *Construction and Building Materials*, 45, 251–261. with permission from Elsevier.)

greater influence on the surface "skin" than on the core of a member, which often "self-cures." (See Chapter 4 for more on these issues.)

This outdated practice derives from an earlier era where mainly labora-tory testing was used to correlate compressive strength with key durability parameters such as abrasion resistance, carbonation and chloride resistance, and soft water attack. However, binders at that time reflected a limited range of mostly "ordinary Portland cement" where the binder chemistry and physical properties were fairly uniform, and comparability could be established across different experimental setups. Strength-durability cor-relations could be reasonably established from laboratory tests in which binders were "simple," concretes were well-cured, and concrete properties were in a limited range. Observations of *in situ* performance and practical experience meant that simplified "rules" could be established, related to limits on mixture proportions and appropriate materials as well as possibly a measure of curing. Specifications thus contained (and mostly still contain) restrictions on maximum w/c ratio, minimum cement content, minimum compressive strength, and so on. (Chapter 5 deals with specifications for durability and should be referred to.)

Modern concrete design and construction demands more sophistica-tion than the simplified approaches of the past. Infrastructure owners are starting to demand "guaranteed service life" or "maintenance-free service life," and commercial developers or infrastructure owners are sometimes required to prove long-term durability of their developments prior to selling them on. Also, abundant evidence exists that many concrete structures are deteriorating at an unacceptable rate resulting in a huge burden on national infrastructure budgets and resources, including physical resources, which is unsustainable (see Chapter 3).

Engineers require practical solutions to concrete durability problems. If compressive strength is no longer adequate for representing durability, the question is, how should engineers approach the durability problem? What practical tools are available for characterizing durability and providing solutions for achieving durability in actual construction? How best can assurances be given to clients and owners on the likely durability performance of a structure?

Answers to these questions are not simple. Engineers mostly work to codes of practice, which are often slow in being updated, so that new knowledge from research and practice takes a long time to find its way into standards. An example of current thinking on durability approaches is given in the *fib* Model Code (2010), and relevant sections are briefly reviewed below.

6.1.3 *fib* Model Code (2010)

The *fib* Model Code (2010) was reviewed comprehensively in Chapter 5 and the reader should consult that chapter for details. The Model Code comprises the following approaches for durability design: (i) deemed to satisfy, (ii) avoidance of deterioration, (iii) partial safety factor, and (iv) probabilistic. The first, that is, "deemed to satisfy," is prescriptive in nature, while the "partial safety factor" and "probabilistic" approaches are more performance-based.

A summary of the Model Code service life design approaches is given in Figure 6.3, for completeness.

The Model Code approach has merit in moving away from current simplistic approaches. For many if not the majority of RC structures, a minimalist approach may be adequate, since many structures are not exposed to severe environments that threaten their longevity. EN 206-1 (2013) has an exposure category of XO, described as "Concrete with reinforcement or embedded metal: Very dry," that is, "Concrete inside buildings with very low air humidity," which represents a large proportion of concrete construction in very mild or benign environments. For such exposures, simple attention to good construction practices, such as good mixture design, compaction, curing, and so on, should ensure adequate durability.

This chapter deals with durability indicators, which are a set of tools to quantify and control concrete durability in construction. Thus, they could be fitted into code approaches in one form or another. However, much of this work remains to be done.

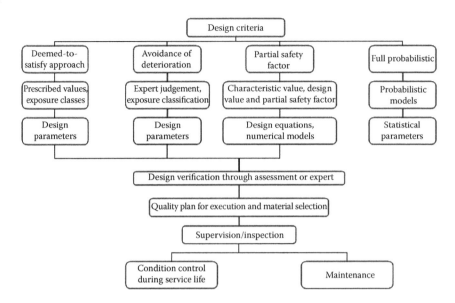

Figure 6.3 Summary of service life design approaches. (Reproduced from ISO 13823, 2008 with the permission of the South African Bureau of Standards [SABS]. The South African Bureau of Standards is the owner of the copyright therein and unauthorized use thereof is prohibited. The standard may be obtained at the following: www.store.sabs.co.za)

6.2 DURABILITY INDICATORS; DURABILITY INDEXES

"Durability indicators" or "durability indexes" carry somewhat different meanings depending on their source and application.* They refer to concrete parameters—mechanical, physical, chemical, electrochemical—that describe a transport property or deterioration mechanism, and thus may be used to characterize the concrete in terms of its "potential" durability. "Potential" durability means the potential for the concrete to be durable in the given environment provided it is properly proportioned with the correct constituents and then cured appropriately *ab initio*. A distinction must be drawn between "potential" durability and actual "as-built" durability, since construction processes such as compaction and curing can result in concrete of a lower quality *in situ* that does not realize its potential for being durable. It is also important actually to measure the as-built properties of concrete *in situ* if assurance is to be obtained as to the long-term performance of the structure.

---

* While these two terms have somewhat different meanings as indicated, for the balance of this chapter they will be used interchangeably and equivalently, and abbreviated where appropriate as "DI."

6.2.1 Philosophy of durability indicators or indexes

Measurement is essential to the achievement of durability in structures. We need relevant durability parameters that can be easily and unambiguously measured, such as permeability, water absorptivity, ionic diffusivity, resistivity, and so on. Two aspects are critical: the purpose for which they are measured, and the ease and speed of measurement so that they can be used as rapid indicators of quality, for example, in quality control schemes in the laboratory or on-site. It is this combination of purpose and ease and speed of measurement that distinguishes whether a parameter can be used as a durability indicator.

A good illustration of the above concept is compressive strength, which has been used successfully for many decades as an "indicator" of concrete quality. Engineers have resolved the issue of whether a concrete structure is "acceptable" from an overall strength perspective by adopting a simple quality control test—the cube or cylinder compression test. The test bears little resemblance to the conditions existing in a real structure: the cube or cylinder is prepared and cured very differently to the concrete in the actual structure and does not experience the same stresses or environmental conditions that a member in the real structure does. Nevertheless, experience has permitted the correlation of compressive strength with structural performance so that structures may be designed for different levels of stress and under different loading systems. In essence, the compressive strength test can be thought of as an "index" test, which characterizes the intrinsic potential of the material to resist applied stresses. By experience, this "index" is associated with acceptable performance and has generally worked very well.

Applying this concept to concrete durability, we need various parameters that can serve as "indexes" of the durability of the material or structure. By measuring them in the relatively short term, we can use them as indicators of the likely durability performance of the structure in the long term. They must preferably be fundamental material parameters that relate to deterioration processes, and used to characterize the concrete in terms of its durability performance under the specific deterioration mechanism. In other words, these parameters need to adequately characterize the key material property or properties that govern the durability issue of concern.

Further to illustrate this concept with an example, a pervasive problem with RC durability is corrosion of the reinforcing steel, which amounts to the need for adequate protection to the reinforcing. This protection depends mainly on the quality of the cover layer, requiring a means of measuring parameters that reflect the quality of the cover, which is actually a very thin element in relation to the entire structure. A strength parameter is inadequate for this since it measures only the overall bulk response of the material to stress, which has little relevance to the quality of the surface layer that is affected by many more influences—constructional and environmental—than the core of the member. Parameters and associated tests are needed to measure the properties of the cover layer. Such parameters must be sensitive to all the processes that affect the quality of the cover, such as the critical influence of curing. These processes are linked with transport mechanisms such as gaseous and ionic diffusion, water absorption, and so on. Thus, the chosen parameter(s) acts as a "durability indicator" for this property and is able to give information on the likely resistance of the material to the deterioration mechanism(s).

This concept can be taken over a wide range of durability problems: a series of index tests are needed, each index being linked to a transport mechanism relevant to a particular deterioration process. The criteria for index testing will be covered later, but such tests need to be relatively simple, quick, and accurate in the sense that they properly represent the real durability problem.

The usefulness of indicator or index tests will be assessed ultimately by reference to actual durability performance of structures built using the indexes for quality control purposes—a long-term undertaking. A framework for durability studies is therefore necessary, such as that illustrated in Figure 6.4, which schematizes the various elements that need to be considered in the development of durability approaches, such as early-age material indexing, direct durability testing, and observations of long-term durability performance. It also shows the various interactions and links between these elements so that an integrated approach emerges that can be used in durability design and specification of concrete structures.

The various elements in Figure 6.4 are discussed in the following sections.

6.2.1.1 Material indexing

Durability indicators or indexes comprise quantifiable chemical, physical, or engineering parameters that characterize the concrete. They must be sensitive to materials, processing, and environmental factors such as binder type and presence of SCMs, water/binder ratio, other mix constituents, placing and compaction, type and degree of curing, and so on. The purpose of material indexing is to provide a reproducible "engineering" measure of microstructure and properties relevant to durability at a relatively early age (e.g., 28 days). The indexes allow the material to be placed in an overall matrix of possible material values that depend on the factors given above. Thus, it should be possible to produce concretes of similar durability by a number of different routes: additional curing, lower w/c ratio, choice of a different cement or extender, and so on.

The material indexes should also be capable of being used as "proxies" for long-term durability performance, by being measures that give indirect information on, for example, the expected service life of the structure.

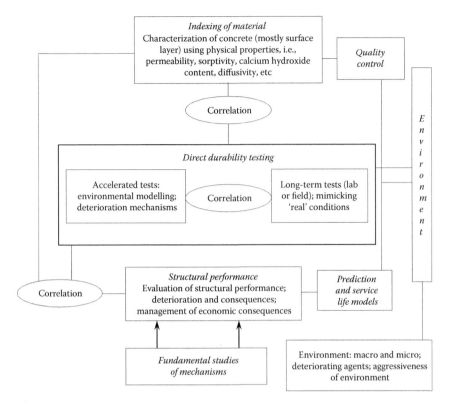

Figure 6.4 Framework for durability studies. (Courtesy of Alexander, M. G. and Ballim, Y. 1993, Experiences with durability testing of concrete: A suggested framework incorporating index parameters and results from accelerated durability tests, in *Proc. 3rd Canadian Symp. on Cement and Concrete*, Ottawa, Canada, August 1993, Nat. Res. Council, Ottawa, Canada, 1993, pp. 248–263.)

Ideally, they should be measured on both laboratory specimens as well as in the actual structure—covered later in terms of material potential versus as-built quality, both of which are important for performance-based approaches to durability design and specification.

6.2.1.2 Direct durability testing

This should comprise a suite of tests suited to a range of durability problems, for example, chloride penetration testing, carbonation testing, and sulfate resistance testing, many of which are already in place in standard test methods. Such tests would include accelerated and long-term evaluations. The need for accelerated testing is obvious in view of the long-time periods involved in concrete deterioration. Nevertheless, it is usually necessary also to do long-term testing, since mechanisms dominant in accelerated tests may differ from those in normal environments. Two other issues regarding direct durability testing are as follows:

- The definition of a suitable measure of deterioration and a threshold or limiting value (e.g., should the extent of sulfate attack be characterized as expansion, mass loss, change in strength and stiffness, and so on?).
- Test methods that as yet may not be accepted standards, or that yield results which are not helpful for characterizing the materials. This prevents useful comparisons between reported results and hampers understanding of mechanisms, or a lack of agreement on what the results of standard tests mean.

6.2.1.3 Structural performance

Observations of long-term structural behavior are also necessary to quantify actual durability performance. This should be done on structures built using indexes or indicators as quality control measures during construction, so that the initial condition of the structure is characterized from a durability perspective. Such studies are time-consuming and laborious and difficult to carry out comprehensively, but without them, the practical usefulness of indexes for achieving durability *in situ* cannot be properly verified.

6.2.1.4 Fundamental mechanistic studies

These studies underpin other aspects discussed above, comprising "materials science" aspects such as microstructural characterization, phase changes, fundamental deterioration mechanisms, and so on.

6.2.1.5 Correlations

Figure 6.4 shows correlations that "link together" the elements discussed above. Some may be obtained from fundamental scientific considerations, while others may require empirical correlations. These correlations permit the key elements to be used in an "integrated" fashion, for example,

- Between indexes and direct durability results, verifying that the indexes are valid representations of deterioration processes or other durability parameters.
- Between indexes (and direct durability results) and actual structural performance—which is the basis for the DI approach; if these correlations do not exist or are unverifiable, then the indexes clearly cannot be used as proxies of durability performance, and they lose their usefulness and power.

Figure 6.4 also indicates the uses to which DIs can be put:

- Quality control—a means of assessing the quality of construction for compliance with a set of criteria.
- Specification criteria—specifying a particular property of concrete, or the quality of a particular concrete element such as the surface layer. Such criteria would typically be reflected by a construction specification in which limits to index values at an appropriate age would be given.
- As a basis for fair payment for the achievement of concrete quality, or alternatively for penalties for non-achievement.
- As a means of predicting the performance of concrete in the design environment.

The criteria for index tests require, *inter alia*, that the tests

- Be site- or laboratory-applicable (site-applicable could involve retrieval of small core specimens from the structure or test element for laboratory testing).
- Be linked to important deterioration and transport mechanisms and have a theoretical basis.
- Be quickly and easily performed without undue demand on equipment or operator skill.
- Have sufficiently low statistical variability to be reliable.
- Avoid difficult or complicated specimen preparation, with uniform preconditioning to ensure standardized testing.
- Be able to be conducted at relatively early ages.

6.2.2 Need for an integrated scheme

The framework diagram in Figure 6.4 shows that material indexing is the key aspect of the scheme: material indexes are used to characterize the potential durability, to carry out quality control, to relate to direct durability measures and deterioration processes, and to link with long-term structural performance. These aspects taken together represent an "integrated scheme," in which measured durability indexes representing the material quality in prequalification tests and trials and also in the as-built structure are linked to the construction specifications for quality control purposes and to predictive service life models (SLMs) used in design. Such a scheme allows complete integration and consistency between design, specification, and actual construction quality.

Up to now, approaches for rational durability design have been limited and not always easy to implement. Examples are DuraCrete in Europe (DuraCrete, 1998) and Life-365 (2005) in the United States (see Chapter 5 for a more detailed description of these and several other similar models). While useful, they tend to be location-specific and only partially represent an integrated approach to durability design, which needs the elements of measurable durability parameters coupled with SLMs. Linked to durability design is also the need for a construction specification to ensure that the design assumptions regarding the concrete quality and composition have indeed been achieved.

6.2.3 "Material potential" versus "as-built quality"

The (legal) responsibility for producing a quality structure normally lies with the constructor. In reality, however, quality and durability of the final structure is the responsibility of both the concrete material supplier (e.g., ready-mix supplier) and the on-site constructor. Therefore, a two-level approach is required, drawing a distinction between material potential and as-built quality. The former refers to the potential of the material to be durable (i.e., what can be achieved), while the latter refers to the durability of the actual structure *in situ* (i.e., what actually is achieved). Durability performance will depend on these two separate but related aspects and it is preferable to allow for them in design and construction, by devising a scheme for acceptance of the as-supplied material so that the concrete supplier can have confidence in the potential quality of the material. This requires the concrete supplier to test samples from the production facility or from the RMC truck using compacted and fully cured specimens that are unaffected by construction effects, in the same way as for strength specimens. The as-built quality is then assessed by sampling the actual structure at some time after construction (say 28 days), by a representative sampling scheme. (It is also possible to construct representative test elements such as panels that "mimic" the identical materials and construction processes in the "real" structure, and test these provided they represent the as-built situation as closely as possible.) If the as-built quality is deficient, the specification should have an internal acceptance scheme that is able to distinguish whether the deficiency arises from the as-supplied material or the manner in which it was processed by the constructor.

As-built durability performance is typically more sensitive to construction processes than strength performance, for example, in the influence on the cover zone where transport of chloride ions and carbonation occur. Except in cases of gross undercompaction, the internal concrete core, which primarily provides strength, is much less affected by most construction processes.

A scheme for implementing the two-level approach discussed above is given in Alexander et al. (2008) and is summarized in the following.

6.2.3.1 Differences between material potential and as-built quality

The effect of site processes on concrete quality can be twofold: a reduction in average quality and increased variability. Thus, the absolute differences in average values between the as-built and material potential specimens, as well as their variances, are required in order to establish the margin between material potential and as-built values. This is illustrated in Figure 6.5a. This has not been extensively studied; however, Gouws et al. (2001) studied the quality of as-built structures and associated site-cast cubes from nominally identical concrete batches, using the South African DI tests. Regarding average values, the results were mixed. As-built values were generally worse, but occasionally better than the laboratory specimens. The occasional reversals were attributed to the methods of finishing, for example, when considerable densification was given to the surfaces of well-cured ground slabs. Considerably more data are required for both laboratory specimens ("material potential") and site-derived specimens ("as-built quality") to quantify typical margins and variances. Table 6.1 gives reported variances

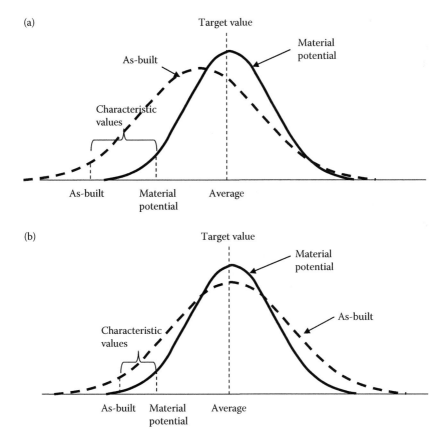

Figure 6.5 Conceptual relationships between material potential and as-built test distributions for a typical durability index test (higher values represent better quality): (a) lower as-built average values; (b) equal average values for material potential and as-built quality.

Table 6.1 Single-operator coefficients of variation

			Oxygen permeability (%)	
			COV for:	
Condition	Chloride conductivity (%)	Water sorptivity (%)	OPI	k
Laboratory (material potential)	5	7	1	23
As-built	14	13	2	50

Source: Reproduced from Gouws, S. M., Alexander, M. G., and Maritz, G., 2001, Use of durability index tests for the assessment and control of concrete quality on site, *Concrete Beton*, 98, 5–16, with permission from the Concrete Society of Southern Africa, CSSA.

Note: k is the D'Arcy coefficient of permeability in m/s units; OPI is the negative log of k.

for laboratory (material potential) and as-built conditions from the study by Gouws et al. (2001).

If differences in average performance between laboratory and site-derived specimens are neglected, a simpler scheme results, as shown in Figure 6.5b. Provided the variances in testing for the different sets of specimens are known, it is possible to quantify the margins between the characteristic values for material potential and as-built quality, as seen in Figure 6.5b. This permits a concrete producer to manufacture concrete to a certain "quality," which will generally be higher than the final quality required in the as-built structure. Provided the quality is measured at both the concrete production stage and in the as-built structure, it is possible to assign responsibilities to the various parties and to determine where the locus of possible poor quality is located. (For details on a scheme for evaluating and implementing margins as described, see Alexander et al., 2008.)

In most cases, *in situ* testing to determine as-built quality is difficult. Better *in situ* tests are required to allow easy and reliable measures of as-built quality. Consequently, construction schemes that require durability testing in the specifications may call for only prequalification testing, that is, for the concrete producer to show that the concrete to be supplied will meet the necessary durability criteria, by way of specified indicator (or other appropriate) values of the concrete.

Much work is still needed on aspects such as "margins" between potential and as-built values—both characteristic and average—influence of site practices on as-built values, conformity and acceptance schemes, practical problems around site testing, for example, action in the event of unacceptable test results, and site targets versus absolute limits.

In summary, durability indexes or durability indicators are measurable material parameters that can be used in durability design and specification. Durability is a "property" to be paid for, just as strength is. Concrete is an "engineered" material in which the requisite properties are provided by design. Each desired property needs to be specified and appropriately paid for. The cost of concrete structures is not likely to dramatically escalate due to such payment for durability. Life-cycle costing almost always proves that paying for durability initially is an investment that is best in the long run.

6.3 MATERIAL PARAMETERS SUITABLE AS DURABILITY INDICATORS

The criteria for parameters suitable for use as DIs given earlier were that they should be site- and laboratory-applicable, quick and relatively easily performed, linked to governing deterioration and transport mechanisms with a theoretical basis, and with sufficiently low statistical variability. A number of these parameters will entail transport properties of concrete. These are covered in Chapters 4 and 7, but are briefly elaborated here, for completeness. The following sets of parameters can be identified as suitable durability indicators, although it is not an exhaustive list.

6.3.1 Physical parameters (physical microstructure of the material)

6.3.1.1 Permeability to liquids and/or gases

Permeation is the movement of fluids through a porous material under an applied pressure while the pores are saturated with the fluid. Concrete permeability depends on its microstructure, moisture condition, and the characteristics of the permeating fluid, with dense microstructures having limited pore interconnectivity and exhibiting low permeability.

Permeation as a mechanism, as well as testing for permeability, is covered in Chapter 7 and will not be further discussed here.

6.3.1.2 Water absorption and sorptivity

Absorption (or sorption) is the uptake of liquids into a solid by capillary suction. It is measured by parameters such as bulk absorption, or sorptivity S, defined as the rate of advance of a wetting front in a dry or partially saturated porous medium:

$$S = \frac{\Delta M_t}{t^{1/2}} \left[\frac{d}{M_{sat} - M_0} \right] \tag{6.1}$$

where $\Delta M_t/t^{1/2}$ is the slope of the straight line produced when the mass of water absorbed is plotted against the square root of time; d is sample thickness; and M_{sat} and M_0 are the saturated mass and dry mass of concrete, respectively.

Absorption is influenced by the larger capillaries and their degree of continuity. Absorption is usually regarded as a bulk material property and measured on cubes or cylinders, while sorptivity is measured on a concrete surface exposed to water. Sorptivity is very sensitive to hydration of the outer surface of concrete, which is influenced by curing and finishing. Sorptivity can be measured either on lab specimens (e.g., ASTM C1585-13 or the South African water sorptivity index (WSI) test [UCT, 2009]) or *in situ* using methods such as the initial surface absorption test (BS 1881: Part 5, 1970), which however has been withdrawn as a standard but still has value, or the Autoclam test developed by Basheer et al. (1994).

6.3.1.3 Porosity; pore spacing parameters

Hardened cement paste is characterized by gel porosity and capillary porosity, with capillary porosity having the greater influence on concrete transport properties due to the significantly larger capillary pore size in relation to gel pores.

Porosity, p, is the internal pore volume as a proportion of the total volume of a solid (Equation 6.2):

$$p = \frac{V_p}{V_T} \tag{6.2}$$

where V_p is the volume of internal pores and V_T is the total volume of the solid.

While the pore volume includes all pores, only the interconnected porosity is measured in laboratory tests, normally by oven drying a sample to constant mass and then saturating it in water.

Porosity influences most concrete properties including strength, permeability, shrinkage, and creep. Regarding durability, porosity profoundly influences all the transport properties of concrete and is therefore a critical parameter that can be used as a durability indicator.

Pore spacing is also a critical parameter, especially in regard to freeze–thaw resistance of concrete. Specifications usually provide requirements on allowable air content, which in effect relates to maximum air bubble spacing to avoid or mitigate freeze–thaw damage. The limits on air content (20 mm aggregate) depend on the severity of exposure conditions and are typically 5%–8% for F-1 exposure and 4%–7% for F-2 exposure (CSA A23.1-04).

6.3.2 Mechanical parameters

The most common mechanical property of concrete and usually its most important property is compressive strength. However, this is not an adequate property for characterizing concrete durability, as argued earlier, being a bulk property. The most common mechanical property related to durability is abrasion resistance (although some regard this as a physical property of concrete).

6.3.2.1 Abrasion resistance

The abrasion resistance of concrete is largely a function of the coarse aggregate abrasion resistance and surface hardness, and the paste–aggregate bond. Abrasion as a durability issue arises where concrete surfaces are exposed to severe abrasive forces such as in pavements and trafficked slabs-on-grade, dam spillways and stilling basins, concrete canals carrying gravel- and silt-laden water, heavily trafficked pedestrian areas, and abrasion-resistant floor toppings. Abrasion parameters typically measured in standard tests include mass loss and depth of wear, and these could be used as durability indicators for abrasion.

6.3.3 Chemical parameters

6.3.3.1 Calcium hydroxide content

Calcium hydroxide content is a measure of the "alkalinity" of the concrete matrix (or more correctly, the calcium oxide content). It is important in assessing the resistance of concrete to carbonation and could be used as a measurable parameter in this respect.

6.3.3.2 Degree of hydration

This parameter can be tricky to measure reliably. It might be useful as a durability indicator when assessing the effectiveness of process and construction variables.

6.3.4 Physicochemical and electrochemical parameters

6.3.4.1 Diffusivity and conductivity

Diffusivity and conductivity refer to the transport of ionic and molecular species through concrete due to a concentration difference or a potential difference, respectively. (Diffusion may also include diffusion of liquids and gases.) Conduction, sometimes called migration, arises from an electrical potential difference and is an electrochemical process. (Migration is also referred to as accelerated diffusion, electrodiffusion, or conduction.) The governing equations are Fick's first law (for steady-state diffusion) and the Nernst–Planck equation (for conduction), which give rise to the following material parameters that can be used as durability indicators:

D_f = diffusion coefficient
σ = conductivity

These two parameters are related through the diffusivity relationship for transport of ionic species:

$$Q = \frac{D_s}{D_o} = \frac{\sigma}{\sigma_o} \tag{6.3}$$

where Q is the diffusivity ratio, σ is the conductivity of concrete (S/m), σ_o is the conductivity of the pore solution (S/m), D_s is the steady-state diffusivity of ionic species through concrete (m²/s), and D_o is the diffusivity of ionic species in the equivalent pore solution (m²/s).

Diffusion in concrete is complicated by several factors: interaction between the diffusant and the matrix (e.g., chloride diffusion), concrete diffusion coefficient reducing with age, and surface concentration of the diffusing species also changing with age. These violate the basic assumptions of the governing equations. Nevertheless, apparent or effective diffusion coefficients or conductivity coefficients are defined and measured, and these are of value as durability indicators provided their limitations are borne in mind.

Resistivity is the inverse of conductivity. It is a very useful electrochemical parameter and is discussed in the following section.

6.3.4.2 Resistivity

Electrical resistivity is an important physicoelectrochemical property of concrete that affects a variety of applications. It is a geometry-independent material property describing the electrical resistance of a material and is fundamentally related to the penetrability of fluids and diffusivity of ions through porous materials such as concrete.

Despite the difficulty in predicting resistivity from knowledge of concrete's constituents and the environment, it is relatively easy to measure, typically using a four-point Wenner probe instrument (Millard et al., 1990), illustrated in Figure 6.6. It is often used as an indirect measure of the probability of corrosion initiation and corrosion rate in RC.

As mentioned, resistivity and conductivity are inverse properties and can be measured in similar ways. However, the tendency is to use resistivity as a measure of corrosion resistance for RC while conductivity is sometimes used as a proxy for chloride diffusion resistance. This is covered later in the section on chloride conductivity testing (Section 6.5.1.7.1).

6.3.4.3 Electrical migration

The chloride migration coefficient is a parameter measured in a migration test. It is obtained from the chloride penetration depth in a voltage-accelerated test (NT Build 492, 1999). This parameter is also used in service life prediction with the DuraCrete approach mentioned earlier and covered in more detail later in this chapter.

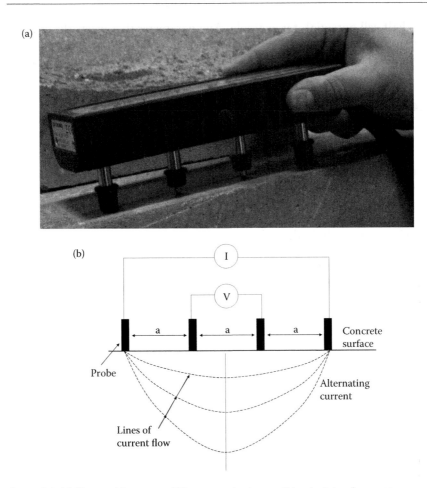

Figure 6.6 (a) Photo of four-point Wenner probe in use; (b) principle of operation.

6.3.4.4 Rapid chloride permeability test

The rapid chloride permeability test (RCPT) (ASTM C1202, 2012) measures the charge passed (coulombs) through a concrete cylinder under a 60-V potential difference. It has existed for several decades and is widely used in durability specifications. It has also been criticized for a lack of a fundamental basis. Nevertheless, as is often the case with even an imperfect test, it has found favor in practice and has been used successfully as a durability indicator for chloride resistance of concrete. It is covered in more detail in Chapter 7.

6.3.5 Other parameters

A full range of possible parameters that could be used as durability indicators is beyond the scope of this chapter, but could include cover depth and half-cell potential, for example. Any parameter that meets the criteria for DIs set out earlier could be considered for possible use. Many of the parameters mentioned above, and others, are considered in Section 6.5.

6.4 PERFORMANCE-BASED DURABILITY DESIGN AND SPECIFICATION: LINK WITH DURABILITY INDICATORS

A particular feature of durability indicators is that they can be incorporated into performance-based durability design and specification, since their philosophy and development was aimed at improving the quality of concrete construction by moving from a prescriptive to a performance-based approach. Indeed, if durability indicators are to find widespread engineering application, they must be incorporated into performance-based durability design and specification. This section outlines such possibilities. (Performance-based approaches to durability design are covered in detail in Chapter 5, in regard to various international specifications.)

6.4.1 Prescriptive versus performance-based design and specification

Design for durability should ensure that concrete structures remain functional for their service life, considering environmental exposure and deterioration, penetrability of concrete influenced by mix constituents and construction processes, cover depth, and service life. The prescriptive approach sets limiting values for concrete mix compositions (w/b ratio, minimum binder content, binder type) and for concrete strength grade and provides guidelines on execution of construction. These limiting values are based on laboratory and field tests, empirical relationships, and past experience. However, the parameters specified are difficult, if not impossible, to verify *in situ*, they cannot account for rapid developments in new materials and processes, they stifle innovation, and perhaps most tellingly, despite the prescriptions becoming ever more onerous with time, the quality of concrete construction has in general not shown corresponding improvement.

6.4.1.1 Performance-based approach to durability design

Somerville (1997) proposed a quantitative approach to the durability design of structures, similar to that adopted in structural design. Somerville described the performance-based approach as an "engineering approach," which should be based on the consideration of five aspects: quantification of "loads" from the environment and the predominant deterioration mechanisms; performance criteria of a structure, for example, service life; prediction models that consider the type and rate of deterioration; factors of safety that consider variability in environmental loads and precision of models; and specifications and quality assurance systems that verify compliance with the required performance. Figure 6.7 provides a schematic of the performance-based approach according to Somerville.

Performance-based durability design can be implemented in a number of ways, for example, a limit-state methodology according to ISO 13823-1 (2008). However, in any methodology, it is critical that the performance parameters and criteria for the structure be explicitly described and a scheme be set up to verify these parameters in practice and ensure the criteria are met. According to the U.S. National Ready Mixed Concrete Association (NRMCA), "A performance specification is a set of instructions that outlines the functional requirements for hardened concrete depending on the application. The instructions should be clear, achievable, measurable, and enforceable. Performance specifications should avoid requirements for means and methods and should avoid limitations on the ingredients or proportions of the concrete mixture" (see Lobo et al., 2005). These functional requirements may be related to strength, durability, or dimensional stability.

Features of performance specifications are therefore the following: (i) the functional requirements should be clearly defined to ensure that the parties involved in their implementation (concrete producers and contractors) do not interpret them differently; (ii) compositional and proportioning requirements of the concrete mix are not given but the concrete producer and constructor work together in the design of the concrete mix, allowing for flexibility in materials selection and ensuring that the concrete produced

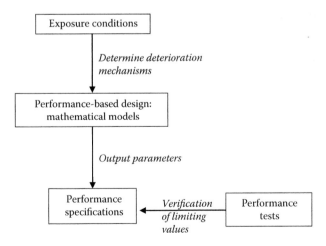

Figure 6.7 Schematic representation of the performance-based approach. (Courtesy of Somerville, G., 1997, Engineering design and service life: A framework for the future, in *Prediction of Concrete Durability: Proceedings of STATS 21st Anniversary Conference*, J. A. N. Glanville (ed.), E & FN SPON, London, UK, pp. 58–76.)

and supplied will meet the requirements of workability; (iii) verification of compliance using tests that are reliable, repeatable, accurate, and applicable on-site; and (iv) a means to enforce compliance with the specifications, for example, penalties when specifications are not met.

The performance-based approach provides a rational means of durability design, since the factors that influence durability are verifiable—typically concrete cover quality and penetrability, and thickness (for avoidance of corrosion). The benefits are several. First, the application of SLMs permits design and construction of RC structures with an owner-specified service life that is notionally quantifiable. Second, as the material compositional requirements are not specified, concrete producers have flexibility to select materials that enable them to take advantage of locally available or more appropriate materials. Third, it encourages innovation. Also, the verification of as-built properties of the structure enforces stricter quality control of construction practices to ensure compliance with performance specifications.

6.4.2 Durability indicators and performance-based durability design

The performance-based approach to durability design involves selection and verification of concrete durability parameters. For example, deterioration due to alkali–aggregate reaction might require limiting the available alkalis. Since the essence of performance-based approaches is measurement and verification, it is easy to see that durability indicators are a powerful tool in achieving desired performance, if they represent the deterioration mechanisms expected during the service life of a structure and if measurable on prequalification mixes (and preferably on the actual structure).

Later, this chapter covers international examples where durability indicators have been implemented in practice. However, to illustrate the use of DIs in a performance-based approach, the South African experience will be summarized here.

6.4.2.1 South African experience: Durability indexes and a performance-based approach

The South African performance-based approach is based on durability index tests described later in this chapter. These characterize near-surface properties of concrete and measure potential resistance against fluid and ionic transport mechanisms that initiate and sustain the corrosion process. The tests were developed as simple and quick experimental procedures that are easy to operate, applicable on both site elements and in laboratories, requiring minimum sample preparation, and conducted at an early age (typically 28 days), making them suitable for quality control (Alexander and Beushausen, 2009).

6.4.2.1.1 Durability index-based performance design

The oxygen permeability index (OPI) test is applicable in SLMs for carbonation-induced corrosion. Correlations have been established between early-age OPI values and carbonation in concrete exposed to different exposure conditions in the long term. For chlorides, good correlation was established between chloride conductivity index (CCI) values (modified to account for binding effects and maturity) and diffusion coefficients of actual structures and site-based exposure samples. A pragmatic model using the CCI for use in marine environments has also been developed, which considers material properties (different binders), construction processes (curing methods), environmental conditions, and the effect these have on chloride ingress (Alexander et al., 2001, 2008).

A further development has been a framework for a probabilistic approach for the chloride-induced corrosion model considering the uncertainties in the physical, statistical, and model aspects (Muigai et al., 2009, 2012). A probabilistic approach is beneficial as it refines the prediction model and increases its reliability.

6.4.2.1.2 Durability index-based performance specifications

The limiting DI values used in performance specifications consider a matrix of factors in a "multifactor" approach: environmental exposure conditions (adapted from EN 206 (2013) and modified for South African conditions), cover depth, required service life, and material properties (binder types). The limiting DI specification values can be determined using either (a) a rigorous approach where limiting values are determined by use of the relevant service life prediction model; this is suitable for durability-critical structures and allows flexibility in selecting the critical parameters or (b) a "deemed-to-satisfy" approach where for typical construction scenarios, limiting values are determined using SLMs and are tabulated. This approach is simple and suitable for routine application (Alexander et al., 2008). As an example of the latter approach, consider Table 6.2, which provides an illustrative summary of the limiting DI (CCI) values for marine environments with service

Table 6.2 Maximum chloride conductivity values (mS/cm) in the as-built structure; 100-year life, 50 mm cover (illustrative values only)

Max. CCI values (mS/cm)—as-built structure (100 years, 50 mm cover)			
EN 206-1 class	70:30 CEM I: fly ash	50:50 CEM I: GGBS	90:10 CEM I: CSF
XS1	2.50	2.80	0.80
XS2	2.15	2.30	0.50
XS3	1.10	1.35	0.35
XS3 (40 mm cover)	0.75	0.95	0.25

life of 100 years and cover depth of 50 mm. Limiting DI values depend on the binder type and exposure condition. The last row in the table also illustrates that, if the cover is reduced from 50 to 40 mm, the required CCI values reduce to reflect a better quality concrete. Thus, there is a "trade-off" between material quality (DI value) and quantity (cover thickness) that allows optimization of mix and construction.

A two-level approach has also been developed for DI-based specifications where the responsibility for production of durable concrete is shared by the material supplier and contractor. The material supplier selects appropriate constituent materials and proportions, which determines the "material potential" of the concrete. The constructor is then required to process the concrete on-site (placing, curing, and finishing) to ensure that it attains the limiting DI value in the specifications. The measured DI value represents the finished ("as-built") quality of a structure.

6.5 USE OF DURABILITY INDICATORS: CONTEMPORARY APPROACHES AND EXAMPLES OF IMPLEMENTATION

6.5.1 Approaches in different countries

Durability indicators have been differently applied, depending on local or regional imperatives. This section reviews the scope of contemporary approaches in different countries in the context of performance-based methodologies. The philosophy underlying the applications as well as major durability problems has dictated the directions taken. It is understandable that for this emerging technology, there will be points of divergence. Nevertheless, similarities exist and with time, convergence may evolve giving rise to more universal standards.

This review considers approaches and examples in Canada, France, the Netherlands, Norway, Spain, Switzerland, and South Africa, regarding parameters and test methods and associated SLMs. A summary of the durability index performance-based approach in these countries is provided in Table 6.3.

6.5.1.1 Canada

The Canadian approach to performance-based design is given in CSA A23.1-1, which defines performance specifications as a method of specifying a construction product in which a final outcome is given in mandatory language, such that the performance requirements can be measured by accepted industry standards and methods. The contractors, material suppliers, and manufacturers are given liberty to choose processes, materials, and activities to use in construction. The provisions considered are class of exposure (e.g., exposure to chlorides, freeze–thaw, aggressive

Table 6.3 Summary of the durability indicator/index performance-based approach in various countries

Country	DI parameter	Service life model	DI test method	Practical application
Canada	Chloride ion penetrability	None	ASTM C1202 Chloride penetration test	ERS in highway projects (MTO)[a] Construction of an underground mass transit station
France	Chloride diffusion coefficient—effective and apparent	LCPC	Chloride diffusion–migration and diffusion tests	Evaluation of *in situ* state of the Vasco da Gama bridge
	• Apparent gas permeability • Liquid water permeability • Initial Ca(OH)$_2$ content • Water-accessible porosity	Empirical models	Air and water permeability	
Netherlands	Chloride ion penetration	DuraCrete	NT Build 492, rapid chloride migration test	Prequalification of concrete mixtures for service life design of structural concrete (NT Build 492) and production control (TEM)
		Probability-based durability design	Two electrode method (TEM)	
Norway	Chloride diffusivity	DURACON	NT Build 492, rapid chloride migration test	Durability design, quality assurance and operation of major concrete structures in severe environments
		• Probability-based durability design	Two electrode method (TEM)	
Spain	Electrical resistivity	• Resistivity-based model • LIFEPRED	Two-point or Wenner four-point resistivity test	Design of concrete mixtures for new Panama Canal

(Continued)

Table 6.3 (Continued) Summary of the durability indicator/index performance-based approach in various countries

Country	DI parameter	Service life model	DI test method	Practical application
Switzerland (SN 505 262/1:2013)	Chloride migration		Max limits: SN 505 262/1-B (NT Build 492)	• Compliance control by concrete producer
	Accelerated carbonation		Max limits: SN 505 262/1-I	• Compliance control by concrete producer
	Air permeability on site		Max limits: SN 505 262/1-E (Torrent's *kT*)	• Compliance control of site concrete end-product
Switzerland/ Japan/China/ Argentina	Cover depth and air permeability on site	Ref-Exp and other models	Covermeters + air-permeability *kT*	• QC and reproducibility test in Swiss tunnel and bridge[b] • QC of precast elements in Hong Kong–Zhuhai–Macau link[c] • Service life prediction of Port of Miami Tunnel[d] • Service life prediction of Tokyo's Museum of W. Art[e]
South Africa	Oxygen permeability	Chloride- and carbonation-induced corrosion initiation models	Oxygen permeability index	Durability index performance-based specifications for a large-scale infrastructure project
	Water sorptivity		Water sorptivity index	
	Chloride conductivity		Chloride conductivity index	

[a] MTO, Ministry of Transport Ontario.
[b] Jacobs et al. (2009).
[c] Wang et al. (2014a,b) and Torrent (2015).
[d] Torrent et al. (2013).
[e] Imamoto et al. (2012).

acids and sulfates), maximum water/cementing materials ratio, minimum specified compressive strength and age at test, air content, curing type, cement restrictions (only for sulfate exposure conditions), and maximum limit on ASTM C1202 (Electrical Indication of Concrete's Ability to Resist Chloride Ion Penetration—RCPT) chloride ion penetration (Hooton et al., 2005).

The definition of performance should be clear to all parties involved in construction since performance can be interpreted differently. In developing the Canadian performance specifications, key concepts were the following:

1. Ability of the specification writer to discern the performance requirements appropriate to the owner's intended use of the concrete structure
2. The need for clear, unambiguous, and quantitative performance characteristics so that performance can be evaluated
3. Availability of reliable, repeatable test methods that evaluate required performance characteristics, as well as compliance limits that take into account the inherent variability of each test method
4. Ability of the concrete producer–contractor team to choose combinations of materials, mixtures, and construction techniques to meet the required characteristics so that projects can be planned and bid upon, risks and costs assessed, and materials and construction operations adjusted to comply with performance requirements

The standard defines responsibilities of parties involved for a prescriptive versus performance specification. The responsibility of the concrete supplier ends at point of placement but the supplier should work with the contractor to ensure the owner's requirements are achieved.

Hooton and Bickley (2012) outline stages at which performance tests can be carried out before or during construction:

1. Prequalification to provide a mixture that, when placed under defined conditions, can meet the specification requirements
2. Quality control to ensure that materials supplied meet the specification, concrete supplied is equivalent to that which was prequalified, and that prequalified placing practices are being followed
3. In-place testing using nondestructive testing and/or extracting cores from the structure to ensure that concrete supplied and placement methods meet owner-defined performance levels

The interested reader can consult the latest Canadian Standard (CSA A23.1, 2014; CSA A23.2, 2014) for examples of performance specification clauses.

The performance specifications have been applied in several highway agency projects and are often referred to as end-result specifications (ERSs), in which contractors are paid on the basis of consistently meeting specified performance requirements, assessed using in-place testing. The Ministry of Transportation in Ontario (MTO) has adopted ERS and has implemented bonus and penalty payments, thus providing positive incentive for contractors. The contractor is required to obtain cores at random where the number of cores depends on the quantity of concrete and type of structural element. A comparative study of test results from the use of both prescriptive specifications and ERSs was undertaken for a period of 10 years by MTO. The ERS test results were more consistent with lower variability than those for which prescriptive specifications were used (Hooton et al., 2005).

Initially, the specified ASTM C1202 test limits were used only for prequalification of concrete mixtures, but this has been modified to include a maximum single value limit as well as average value coulomb limit when the test is to be used for acceptance testing (CSA A23.1). This allows for consideration of test variability, similar to the approach used for statistical analysis of strength results (Hooton and Bickley, 2012). For example, for C-1 exposure conditions (exposure to freezing in a saturated condition with de-icer salts), the limit on chloride penetration is 1500 coulombs at 91 days, and for acceptance during construction, a single test value is allowed up to 1750 coulombs as long as the average value is below 1500 coulombs at 91 days. (Prior to the 2014 revision of CSA A23.1, these limits were at 56 days.)

Hooton and Bickley (2014) provide an example of the application of performance specifications for a large-scale project where the owner's requirements were durable, leak-proof mass concrete for an underground mass transit station situated below the water table. The owner's specification requirements were clearly defined with quantitative measures for compressive strength, minimum in-place air content, maximum RCPT index at 56 days, maximum shrinkage, and maximum allowable temperature gradient. The responsibilities of the concrete supplier and contractor were also outlined. The prequalification testing to determine a suitable mix required casting and testing monoliths (minimum dimension of 1 m × 1 m × 1 m in insulated forms, instrumented with thermocouples to determine maximum temperature rise and gradients). The specification also provided details on quality control testing during construction and frequency of testing.

6.5.1.2 France

The performance-based design approach allows design of concrete mixtures capable of resisting certain types of degradation (e.g., reinforcement corrosion) for a required service life under given environmental conditions. The approach can be used for durability prediction at the design stage of a structure or for evaluation of residual durability of in-service structures. The durability assessment consists firstly of a qualitative or quantitative evaluation of the "potential" durability by measuring various properties (i.e., durability indicators such as apparent chloride diffusion) on laboratory samples. The performance of various concrete mixtures can thereby be compared and ranked in classes ranging from very low to very high "potential" durability. These classes are then used by design engineers in selecting concrete mixes or optimizing the mix design for a given RC structure (Baroghel-Bouny, 2004).

The models used in the prediction of service life vary in sophistication and are selected depending on the importance of the structure (thus determining the accuracy required for prediction) and the availability of data (Baroghel-Bouny et al., 2006). The high-level models are based on well-identified physical and chemical mechanisms, which include moisture transport (realistic wetting–drying cycles), and take into account the microstructural changes induced by the degradation processes. Where input data are few, simple empirical models can be used, for example, applying Fick's second law of diffusion. A physical model, "LCPC," is used in predicting service life or for evaluating residual life of existing structures. The accuracy of output from the LCPC model varies depending on the service life requirements of the structure and thus the level selected. Level 1 is a simple model that considers the diffusion of chlorides, Level 2 is a multispecies model, and Level 3 is an advanced model accounting for up to seven ionic species in the pore solution. The input parameters in the model are "durability indicators," for example, diffusion coefficients, and the output is the monitoring parameters. For carbonation-induced corrosion, an empirical model from LEO-EDF software (Baroghel-Bouny et al., 2006), an analytical physicochemical model developed by Papadakis et al. (1991), and a numerical physicochemical model (LCPC model) are considered.

The durability evaluation for materials involves measuring various properties, that is, durability indicators, by means of accelerated tests on laboratory samples water-cured for 90 days. Baroghel-Bouny (2004) notes that a single parameter is not sufficient to characterize the behavior of concrete because of various driving forces that are involved in the transport of aggressive species, for example, concentration gradient or total pressure gradient, and the complex physical and chemical processes that take place. A set of indicators that considers basic properties, referred to as universal durability indicators, are applicable to many degradation processes and includes porosity accessible to water (%), effective and apparent chloride diffusion coefficient (measured using either migration or diffusion tests), apparent gas permeability, liquid water permeability, and $Ca(OH)_2$ content (by mass of cement). These universal indicators are complemented by specific indicators that are selected on the basis of environmental conditions and required service life of the structure; for example, for chloride-induced corrosion, these include effective and apparent chloride diffusion coefficients in saturated conditions, porosity accessible to water, and electrical resistivity. In addition to the durability indicators, monitoring parameters

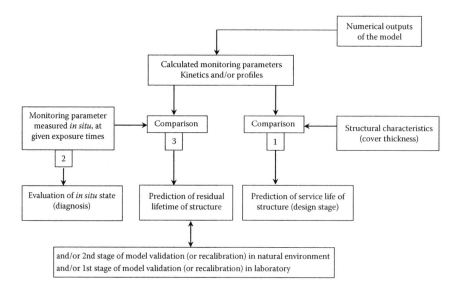

Figure 6.8 Implementation of the predictive approach based on durability indicators and showing the role of monitoring parameters in service life prediction, diagnosis, and prediction of residual lifetime of a structure. (After Baroghel-Bouny, V., 2004, Durability indicators: A basic tool for performance-based evaluation and prediction of RC durability, in *International Seminar on Durability and Lifecycle Evaluation of Concrete Structures*, R. Sato, Y. Fujimoto, and T. Dohi (eds.), University of Hiroshima, Higashi-Hiroshima, Japan, pp. 13–22.)

are used, for example, a simple and rapid colorimetric method to assess kinetics of chloride penetration. The use of monitoring parameters is illustrated in Figure 6.8.

This performance-based approach thus provides a toolkit for the evaluation and prediction of RC durability, with a set of universal and specific durability indicators and monitoring parameters that are applied at the design stage and during the monitoring of existing structures (Baroghel-Bouny, 2002).

6.5.1.3 The Netherlands

Polder and de Rooij (2005) and Polder et al. (2006) reviewed a study in the Netherlands to validate the use of the DuraCrete SLM for RC structures in exposure conditions prone to chloride-induced corrosion. A site investigation that considered six existing structures constructed between 1960 and 1984 was done. The structures were *in situ* or precast, with Portland cement or blast furnace slag cement. Tests carried out on a 1 m × 1 m test area included measurement of resistivity and cover depth, and extraction of six cores to determine chloride profiles. The DuraCrete model was used to compare model predictions of the chloride profile with those measured. The predicted values were observed to be close to measured values except for the outer 1-cm surface layer where there were large differences. To correct for this, the model was modified by adjusting the formulation for the time dependency of the diffusion coefficient, by allowing for mean temperature during the year, and by adjusting the aging exponent.

A performance-based approach for chloride-induced corrosion using the DuraCrete model is also provided in van der Wegen et al. (2012). Three options can be considered when selecting the performance-based approach: (i) a full probabilistic approach based on specified input parameters that consider mean values and variability, (ii) semi-probabilistic approach that uses a safety margin for the cover depth, and (iii) a range of cover depths adapted to binder type and exposure class. The chloride diffusion coefficient of the concrete is determined using the rapid chloride migration test (RCMT) (NT Build 492, 1999), which has a good linear correlation with diffusion coefficients from pure (immersion) diffusion tests (NT Build 443, 1995). The RCMT has been applied for many concrete mixtures in association with service life design for large infrastructure projects. For regular production control, the two-electrode method (TEM) for measuring resistivity is used, which enables a quick, simple and nondestructive test on any regularly shaped specimen. A correlation has been established between resistivity and chloride transport in concrete. The test samples used for TEM are standard concrete cubes for compressive strength testing after wet curing at an age of 28 days. The performance specifications for a given service life provide a maximum RCMT value, mean cover depth (reinforcing or prestressing steel) for a given binder type and exposure conditions.

Prior to adopting TEM-based control of production concrete in van der Wegen et al. (2012), the approach was applied in the construction of the Green Heart Tunnel, which was designed to have a technical service life of 100 years (de Rooij et al., 2007). The limiting durability parameter values were a minimum chloride diffusivity of 5×10^{-12} m^2/s tested using rapid chloride migration (RCM) at 28 days, minimum cover depth of 45 mm, and blast furnace slag cement with a high slag content, CEM III/B. Many batches of concrete were tested over several years of production by TEM for compliance to a preset limit based on the correlation between TEM and inverse chloride diffusion (tested by RCM on a limited number of batches). All batches tested were found to comply with the TEM limit. The durability performance was validated *in situ* with concrete samples of age between 1 and 3 years. Eight test areas of different structural elements were obtained, which comprised 2×2 to 2×4 m^2 of the concrete surface. For each test area, the cover depth was measured, and 12 cores (100 mm diameter and 250 mm long) were obtained for compressive strength and RCM testing. From the analysis of data, the mean cover depth ranged from 45 to 81 mm, minimum cover depths from 5 to 66 mm, and the standard deviation from 4 to 16 mm. The mean RCM values measured at ages between 1 and 3 years ranged from 1.20×10^{-12} to 2.76×10^{-12} m²/s with low standard deviations, a maximum of 1.54×10^{-12} m²/s. For checking versus the original limit at 28 days, RCM values had to be calculated backwards to 28 days. This was carried out by applying a time exponent of 0.27. Using this "model," 28-day values were about 3.5×10^{-12} m²/s, which was below the maximum value provided in the specifications (5.0×10^{-12} m²/s).

The procedure that was followed for batch "durability performance" control appeared to have been successful. Consequently, it was adopted in the Recommendation for service life design of structural concrete (CUR, 2009; van der Wegen et al., 2012).

6.5.1.4 Norway

According to Gjørv (2014), the durability of concrete structures in severe environments is related not only to design and materials, but also to construction. Thus, many durability problems can be ascribed to lack of proper quality assurance during concrete construction and poorly achieved construction quality. The achieved construction quality of new concrete structures generally shows a high scatter and variability, and during operation of the structures, any weaknesses and deficiencies will soon be revealed whatever durability specifications and materials have been applied. To a certain extent, a probability approach to durability design can accommodate the high scatter and variability. However, a numerical approach alone is insufficient for ensuring durability; greater control and improvements in durability also require the specification of performance-based durability requirements, which can be verified and controlled during concrete construction in order to achieve quality assurance.

In order to obtain better controlled and increased durability and service life of new major concrete infrastructure in Norway, the probability-based DURACON model has been used as a basis for durability design, quality assurance, and operation of a number of new concrete structures in recent years—see Figure 6.9 (Gjørv, 2014). In principle, the DURACON model is a simplified and modified version of the DuraCrete Model (2000). On the basis of a modified Fick's second law of diffusion and a Monte Carlo simulation, a method is obtained for calculating the probability of corrosion during a certain "service period" for the given structure in the given environment (Ferreira, 2004; Ferreira et al., 2004). Special software (DURACON) is applied, where the following input parameters are needed:

1. *Environmental loading*: chloride loading (C_S), age at chloride loading (t'), and temperature (T)
2. *Concrete quality*: chloride diffusivity (D), time dependence of the chloride diffusivity (α), and critical chloride content (C_{CR})
3. *Concrete cover*: nominal concrete cover (X)

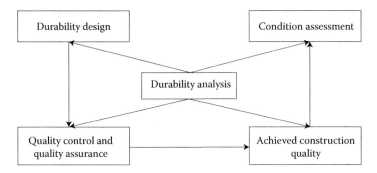

Figure 6.9 DURACON model as a basis for durability design, quality assurance, and operation of new major concrete infrastructure in severe environments. (From PIANC Norway/NAHE, 2009, Durable concrete structures—Part 1: Recommended specifications for new concrete harbor structures, in *Part 2: Practical Guidelines for Durability Design and Concrete Quality Assurance*, 3rd edition, Norwegian Association for Harbor Engineers (NAHE), TEKNA, Oslo, Norway (in Norwegian); reproduced from Alexander, M.G. (ed.), 2016, *Marine Concrete Structures: Design, Durability and Performance*, Woodhead Publishing (Imprint of Elsevier), Cambridge, UK, with permission from Elsevier.)

In the DURACON model, a certain "service period" is specified, before the probability for onset of steel corrosion exceeds an upper serviceability level of 10%, which is in accordance with current standards for reliability of structures. On the basis of the calculations, a combination of concrete quality and concrete cover can be selected, which will meet the specified "service period."

A limitation to the method is that the calculations of corrosion probability (durability analyses) are based on a very simplified diffusion model for a one-dimensional ingress of chlorides and a number of very uncertain input parameters. Therefore, the DURACON model should not be used for prediction of any service life, nor should the "service periods" before the 10% corrosion probability is reached be considered as any real time until start of corrosion. The above "service periods" should rather be considered the result of an engineering assessment of the most important parameters related to the durability of the given structure in the given environment, including the scatter and variability involved. In this way, it is possible to quantify the combined effects of concrete quality and concrete cover for comparison and selection of one of several technical solutions, in order to obtain a best possible durability design.

The procedures and methods for determining the input parameters are discussed in more detail elsewhere (Gjørv, 2014), where a free download link to the DURACON Software is also given. The chloride diffusivity (D) of the given concrete is a very important quality parameter, which for DURACON is based on the RCMT method (NT Build 492, 1999). Since the RCM method is a very strongly accelerated test, the resulting chloride diffusivity should be considered only as a simple relative index, reflecting the density and permeability of the concrete as well as the ion mobility in the pore solution of the concrete. Thus, using the 28-day chloride diffusivity (D_{28}) as an input parameter to the durability design may be compared to using the 28-day compressive strength (f_{28}) as an input parameter to the structural design. The 28-day compressive strength is also a very simple relative index primarily reflecting the compressive strength, and hence the general mechanical properties of the concrete. However, it should be noted that the 28-day chloride diffusivity is a much more sensitive concrete quality parameter than 28-day compressive strength.

This approach to durability design establishes requirements for both concrete quality (D_{28}) and concrete cover (X), which provide the basis for regular quality control and quality assurance during concrete construction. Although the RCM method is very rapid, it is not suitable for ongoing concrete quality control. On the basis of the Nernst–Einstein equation expressing the general relationship between diffusivity and electrical resistivity of any porous material (Atkins and De Paula, 2006), a calibration as in Figure 6.10 is established, relating the two tests. Such a calibration curve must be established for the given concrete before concrete construction starts. Measurements of the electrical resistivity are then carried out as a quick, nondestructive test on the same concrete specimens as are being used for regular quality control of the 28-day compressive strength during concrete construction (Gjørv, 2003). Control measurements of electrical resistivity are carried out immediately before compressive strength testing, by the use of either the TEM or the four-electrode method (Wenner) (Sengul

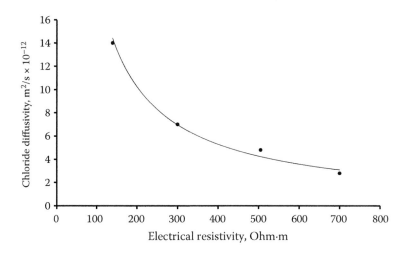

Figure 6.10 Typical calibration curve for control of chloride diffusivity (RCM) based on electrical resistivity measurements.

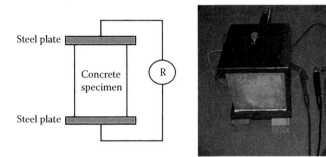

Figure 6.11 TEM for electrical resistivity measurements of concrete.

and Gjørv, 2008; Gjørv, 2014). Since the TEM is more well-defined, better reflecting the bulk properties of the concrete, this method is preferable for quality control during concrete construction (Figure 6.11).

After construction, a further durability analysis is carried out with the obtained average values and standard deviations of the 28-day chloride diffusivity (D_{28}) and the concrete cover (X) as new input parameters. All the other originally assumed input parameters, which may have been difficult to select during design, are kept the same. Therefore, the outcome primarily reflects the results obtained from regular quality control of the 28-day chloride diffusivity and concrete cover during construction, including the observed scatter and variability. Documentation of compliance with the durability specification is therefore obtained.

To provide further documentation of achieved construction quality, some additional durability analyses are carried out. Concrete cores are removed from the structure during construction and tested for development of in-place chloride diffusivity. At the same time, a number of laboratory-cast concrete specimens are also tested for obtaining the potential chloride diffusivity of the concrete (Figure 6.12). With these values of diffusivity and concrete cover obtained during construction as new input parameters, new durability analyses are carried out for documentation of achieved in-place construction quality and potential construction quality of the structure, respectively.

During the operational stage of the structure, further calculations of corrosion probability are carried out as a basis for their regular condition assessment and preventive maintenance. In this case, calculations are based on the apparent chloride diffusivities obtained from the observed rates of chloride ingress in combination with site data on the concrete cover; before this probability of corrosion becomes too high, appropriate protective measure can be implemented.

On the basis of current experience with these procedures for durability design and quality assurance on major new concrete infrastructure, the

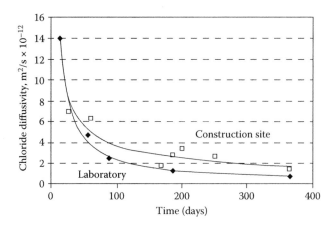

Figure 6.12 Typical development of chloride diffusivity (RCM) on the construction site and in the laboratory during the construction period. (Reproduced from Gjørv, O. E., 2009, *Durability Design of Concrete Structures in Severe Environments*, Taylor & Francis, London, UK, with permission from Taylor & Francis.)

required documentation of achieved construction quality and compliance with the durability specification has clarified the responsibility of the contractor for the quality of the construction process. As a result, improved workmanship giving reduced scatter and variability of achieved construction quality has typically been observed. For the owners of structures, it has been very important to receive documentation of achieved construction quality and compliance with the durability specification before the structures have been formally handed over from the contractors (Gjørv, 2010, 2014, 2015).

6.5.1.5 Spain

Andrade et al. (1993) proposed the use of electrical resistivity to characterize mass transport processes universally in concrete, that is, resistivity can be used to evaluate the chloride diffusion and gas permeation coefficients. Resistivity measurements can be related to porosity, which is then used to determine gas transport processes, that is, permeability. For saturated concrete, the durability parameter considered is the chloride diffusion coefficient, obtained from the Nernst–Einstein equation (Equation 6.4); where concrete is partially moist, the increase in resistivity with respect to saturation is proportional to the decrease in the diffusion coefficient. Thus, Equation 6.4 can be applied to either penetration of chloride (ions) or progress of carbonation (gases). Resistivity provides a fast, easy, and cheap measure of concrete penetrability, making it suitable for use on-site for quality control of new structures. A limitation with the test is that it cannot take into account the binding capacity of concrete and its influence on transport mechanisms.

$$D_{cl} = \left(\frac{RT}{nF^2} \right) \Lambda \tag{6.4}$$

where D_{cl} is the chloride diffusion coefficient (cm²/s), R is the universal gas constant (1/mol °K), T is temperature (°K), n is the number of charges (valence of ion, i.e., chloride), F is Faraday's number (coulomb/eq), and Λ is the equivalent conductivity (cm² ohm⁻¹ eq⁻¹).

Andrade (1993) critically reviewed the RCPT for determining chloride migration. Limitations observed were as follows: (i) the test accounts for total current and not only that corresponding to chloride ion flow, that is, diffusion and migration occur simultaneously and (ii) the high voltage used in the test (60 V) induces heat, which in turn changes the test kinetics. However, the test can still be applied for practical purposes to determine the diffusion coefficients using the Nernst–Plank equation (Equation 6.5) and Nernst–Einstein equation (Equation 6.4). It is assumed that convection does not operate inside the concrete and that diffusion is negligible compared to migration when electrical fields higher than 10 V are used. Therefore, only the migration term is used from which the diffusion coefficient is determined.

$$-J_j(x) = D_j \frac{\partial C_j(x)}{\partial x} + \frac{Z_j F}{RT} D_j C_j \frac{\partial E(x)}{\partial (x)} + C_j V(x) \tag{6.5}$$

$$\text{(Diffusion)} \quad \text{(Migration)} \quad \text{(Convection)}$$

where $J_j(x)$ is the unidirectional flux of species j (mol/cm² s), D_j is the diffusion coefficient of species j (cm²/s), ∂C_j is the variation of concentration (mol/cm³), ∂x is the variation of distance (cm), Z_j is the electrical charge of species j, F is Faraday's number (coulomb/eq), R is the universal gas constant (1/mol °K), T is the absolute temperature (°K), C_j is the bulk concentration of the species j (mol/cm³), ∂E is the variation of potential (V), and V is the artificial or forced velocity of ion (cm/s).

Resistivity also has the advantage of enabling assessment of existing structures through systematic mapping, described in the RILEM recommendation from TC 154-EMC authored by R. Polder (Andrade et al., 2004). In addition to resistivity measurements and RCPT, Andrade et al. (2003) proposed the use of half-cell potential measurements and site determination of the corrosion rate using polarization resistance (Andrade et al., 2004). This can be used for (i) locating corroding rebars and assessing reinforcement corrosion condition during inspection and condition assessment of a structure, (ii) determining positions that can be used to obtain cores for additional testing to obtain corrosion state of rebars, (iii) evaluating the corrosion state of rebars after repair work and determining efficiency and durability of repair work, and (iv) designing anode layout of cathodic protection systems or electrochemical restoration techniques. The data obtained from half-cell potential measurements and the corrosion rate can be represented either statistically using histograms and cumulative frequency distributions, or using half-cell potential maps.

SLMs for the initiation and propagation period of corrosion, based on electrical resistivity, have been developed and are reported in Andrade (2004) and Andrade and d'Andrea (2010)—see, for example, Equation 6.6. The SLM considers the reaction or retarder factor of chlorides (r_{cl}) for different cement types, accounting for chlorides that are immobilized by cement phases through binding, environmental factors (k_{cl,CO_2}) based on exposure classification as in EN 206, and the aging factor (ρ_t). The input parameters in the model are the cement type that determines the value of r_{cl}, exposure class from which a value of k_{cl,CO_2} is obtained, service life, for example, 100 years, cover depth, and the aging factor. From these input parameters, the resistivity is obtained as a corrosion indicator (or durability indicator) that can be used to assess performance of a structure. Alternatively, the cover thickness can be calculated and used for performance specifications where the actual resistivity is known.

$$t_l = \frac{x^2 \rho_{ef}(t/t_0)}{k_{cl,CO_2}} \cdot r_{cl,CO_2} + \frac{P_x(\rho_{ef}(t/t_0)^q \cdot \xi)}{K_{corr}} \qquad (6.6)$$

$$\text{Initiation period} \quad \text{Propagation period}$$

where t_l is the required service life, x is cover thickness (cm), ρ_{ef} is resistivity (Ω cm), t is the time of measurement, t_0 is the time at which initial measurement is taken, k_{cl,CO_2} is the factor of aggressive penetration dependent on exposure class, r_{cl,CO_2} is the reaction factor dependent on cement type, P_x is the limit corrosion attack depth (loss in rebar diameter/pit depth), ξ is the resistivity standardized by its value in saturated conditions, K_{corr} is a constant with a value of 3×104 µA/cm² kΩ cm, and q is the aging factor.

Resistivity and chloride resistance measurements were used to assess the durability of the new Panama Canal (Andrade et al., 2014). The service life was calculated using the resistivity-based model (Andrade and d'Andrea, 2010) and a numerical model LIFEPRED, based on Fick's law. The project specifications require a 100-year service life, which was defined as conformity with 1000 coulombs for electrical charge (ASTM C1202) and application of the SLMs mentioned. The specifications also gave provisions to ensure minimization of heat of hydration-induced cracking in concrete, and cover depth of 10 cm. Concrete mixes were prepared and checked for conformity with the requirements with a Wenner probe.

6.5.1.6 Switzerland

The Swiss Standard SN 505 262/1 (2013) incorporates several DIs for tests on cast specimens, prescribing limiting values for compliance by the concrete producers. Among them are a chloride migration test (similar to NT Build 492, 1999) and an accelerated carbonation test, for chloride- and carbonation-induced corrosion, respectively. In addition, a DI is prescribed for conformity control of the end-product, using the site air-permeability test developed by Torrent (1992), with limiting values for chloride- and carbonation-induced corrosion.

Rules for the application of the Torrent air-permeability test for quality and durability control are provided in Swiss Standard SN 505 262/1 (2013), summarized by Torrent et al. (2012). Limiting coefficients of permeability values, kT, are provided based on the exposure conditions in EN 206-1. To undertake testing, the structure is divided into groups of elements that (i) have the same specified air-permeability values, (ii) are built with concrete belonging to the same EN 206 class, and (iii) are built applying similar concreting practices. The groups are then listed chronologically by date of concreting. For continuous elements, segments concreted on the same day should be identified. The elements within each group are divided into test areas by considering either one test area per 500 m^2 of exposed surface area, or one test area per 3 days of concreting of the elements in the group. The resulting maximum number of test areas should be adopted. For each test area selected, six measurement points are used, which are sampled at random, avoiding excessive closeness to edges and to each other.

The *in situ* concrete should be tested 28 to 90 days after placing. For slow-reacting cements, for example, mixes with fly ash, a minimum test age of 60 days should be considered. Provisions should be taken to avoid testing concrete at very low temperatures or with high degrees of saturation. Moisture contents should be checked using an electrical impedance-based instrument, such as the Concrete Moisture Encounter meter (Tramex, 2014), with an upper moisture limit of 5.5% (by mass). To ensure these moisture conditions are attained, testing should be carried out 3–4 weeks after curing has ended, or 2–5 days or more after the last ingress of water into the concrete by rain or thaw. If the right test conditions cannot be naturally met, test areas should be protected from wetting and allowed to dry until at least one of the above-mentioned conditions for moisture in the structure is fulfilled. Further details on conformity evaluation and acceptance testing are given in the Swiss Standard or in Jacobs et al. (2009) and Torrent and Jacobs (2014).

Given the good correlation found between the coefficient of air-permeability (kT) and other DIs, such as charged passed in ASTM C1202 or carbonation rate (Torrent et al., 2012), a model (Ref-Exp) has been proposed to estimate service life as a function of two key characteristics, both measured on-site: cover depth and kT (Torrent, 2015). The model considers the provisions in EN Standards of maximum water/cement ratio and minimum cover depth, for an expected service life of 50 years. On selection of the maximum water/cement ratio for a given environment, the CEB-FIP Model Code 1990, Equation 2.1-107 is applied to convert w/c ratio into a reference value of gas permeability (kT_{ref}). If the value of kT measured on-site is smaller than kT_{ref} and/or the cover depth (measured in the same place) is larger than the specified value, a service life longer than 50 years is predicted by the model. The Ref-Exp model is based on a *Ref*erence condition (50 year service life) and on *Exp*erimental values measured nondestructively on-site. This approach has been used to verify the service life predictions for large precast submerged elements of the Hong Kong–Zhuhai–Macau link (Wang et al., 2014b; Torrent, 2015).

An air-permeability value is determined by obtaining the geometric mean of the tests for the same lot, which corresponds to the arithmetic mean of the logarithm of permeability data. The geometric mean is preferred to the arithmetic mean, which is very strongly influenced by a few high air-permeability values (Jacobs and Hunkeler, 2006; Jacobs et al., 2009; Wang et al., 2014b).

The application of the Torrent air-permeability test can be done for new construction and for existing structures. Torrent and Imamoto (2015) provide examples of comparative studies of measured air permeability with carbonation depth for Swiss and Japanese structures up to 60 years old. The "Torrent" method to measure air permeability on-site has been adopted in Japan for condition assessment of existing structures (Japan Concrete Institute, 2014).

6.5.1.7 South Africa

In South Africa, the so-called "durability index approach" has been developed over the last two decades (Alexander et al., 1999, 2001, 2008; Alexander and Beushausen, 2009). It commenced with a realization in the early 1990s that the durability of RC construction was becoming a serious issue, especially for owners of major infrastructure such as national freeways, harbors, and water retention structures. The observed durability problems related mostly to premature corrosion of reinforcing steel. Ballim (1993, 1994) gave a major thrust to the initiative by developing two rapid durability-type tests (oxygen permeability and water sorptivity), and Streicher and Alexander (1995) followed this with the development of the rapid chloride conductivity test. The tests have progressed through several round-robin test programs to prove their robustness and reliability and to establish their variability. Also, the South African National Roads Agency Limited (SOC) (SANRAL) has incorporated the approach in their specifications for major national road infrastructure, while other authorities are beginning to do the same. The approach is in use nationally and has been implemented on a large scale (see later).

The premises of the DI approach reflecting the underlying philosophy are as follows:

- The durability of RC structures depends primarily on the quality and quantity (thickness) of the cover layer, that is, its ability to protect the reinforcing steel.
- Improved durability will not be assured in the absence of relevant durability parameter(s), which can be reliably measured.
- A means of characterizing the quality of the concrete cover layer is required, using parameters that reflect deterioration processes and linked with transport mechanisms such as gaseous and ionic diffusion, water absorption, and so on. These parameters are termed "durability indexes." Several indexes will be needed to cover a range of durability problems, each index being linked to a transport mechanism relevant to that particular process.
- The usefulness of index tests will depend on the actual durability performance of structures built using the indexes for quality control purposes.

Durability indexes are quantifiable engineering parameters that characterize the concrete and are sensitive to materials, processing, and environmental factors such as cement type, water/binder ratio, type and degree of curing, and so on. They can be measured either on control specimens (e.g., lab or site-based controls) or on the as-built structure. DIs provide an engineering measure of microstructure and properties critical to RC durability at a relatively early age (e.g., 28 days).

The three index tests are described below (for further details, see literature). They have been shown to fulfill the criteria set out above and are able to characterize the quality of concrete.

6.5.1.7.1 Durability index test methods

These comprise oxygen permeability, chloride conductivity, and water sorptivity index tests. Test equipment and test procedures are described in the literature (Alexander et al., 2001) and basic principles are given below. All three tests use standard 25- to 30-mm-thick, 70-mm-diameter concrete discs cut from cores from lab specimens (e.g., cubes) or actual structural elements, with standard preconditioning at 50°C to ensure uniformly low moisture contents at the start of the test.

The OPI test (also briefly covered in Chapter 7) measures the pressure decay of oxygen passed through a concrete disc in a falling head permeameter (Figure 6.13). The OPI is the negative log of the D'Arcy coefficient of permeability, with common values ranging from 8.5 to 10.5, a higher value

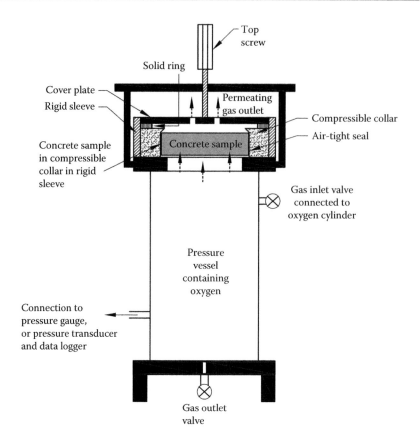

Figure 6.13 Test schematic for the OPI test.

indicating a higher impermeability and thus a concrete of potentially higher quality. The difference between an 8.5 and 10.5 OPI value is substantial because of the log scale. The OPI parameter is also used in a prediction model for carbonation depths.

The water sorptivity test measures the rate of water uptake from a wetting face into a dry sample of concrete due to capillary suction, which depends on pore geometry and saturation. The rate of movement of a wetting front under capillary forces is defined as sorptivity. Several concrete absorption tests exist in which concrete is immersed in water and the total mass of water absorbed is measured as the absorption. These tests reflect concrete porosity but cannot quantify the rate of absorption and do not distinguish between surface and bulk effects.

The SA water sorptivity test is a modified version of Kelham's sorptivity test, developed as a compromise between accuracy and ease of use (Kelham, 1988). Preconditioned samples have the circular edges sealed to ensure unidirectional absorption. The specimen test face is exposed to

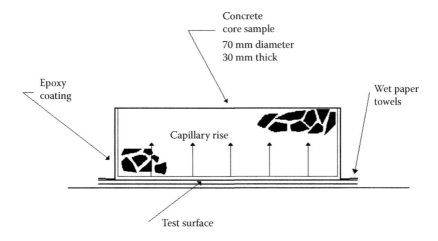

Figure 6.14 Schematic of water sorptivity test.

several layers of saturated paper towels, as shown in Figure 6.14. At regular time intervals, the specimens are removed from the water and the mass of water absorbed is determined. Measurements are stopped before saturation is reached and the concrete is then vacuum-saturated in water to determine the effective porosity.

A linear relationship is observed when the mass of absorbed water is plotted against the square root of time. This relationship, normalized by the porosity, gives the sorptivity.

The chloride conductivity (CCI) test apparatus (Figure 6.15) consists of a two-cell conduction rig with disc samples exposed on either side to a 5 M NaCl solution. The samples are preconditioned before testing to standardize the pore solution (oven-dried at 50°C followed by vacuum saturation and soaking over a 24-hour period in 5 M NaCl solution). The transport of chloride ions is driven by a 10-V potential difference across the specimen, with the chloride conductivity determined from the instantaneous current flowing through the specimen, thus allowing rapid testing under controlled laboratory conditions. Good correlations between 28-day chloride conductivity results and diffusion coefficients after several years' marine exposure exist over a wide range of concretes.

The above tests have been extensively "proved" using inter-laboratory trials (Stanish et al., 2006), although some development is still ongoing. All three have been shown to be sensitive, although in different ways on account of the different transport mechanisms reflected, to materials and processing parameters such as binder type, mix proportions, type and extent of curing, and compaction. The OPI and CCI tests have recently (2015) been accepted as South African National Standards (SANS 3001-CO3-1, 2015; SANS 3001-CO3-2, 2015; SANS 3001-CO3-3, 2015).

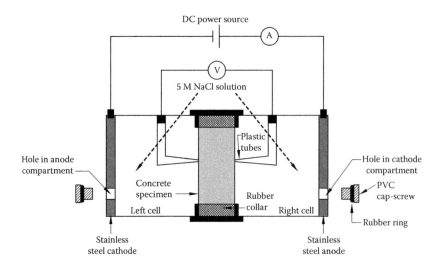

Figure 6.15 Schematic test setup for the chloride conductivity test.

6.5.1.7.2 Application of the durability index approach

The tests described above, when incorporated into the broader philosophical approach to achieving durability previously discussed, can be applied in several ways.

6.5.1.7.2.1 QUALITY CONTROL OF SITE CONCRETE

The sensitivity of the index tests to material and constructional effects makes them suitable for site quality control. Since the different tests measure distinct transport mechanisms, their suitability depends on the property being considered. Permeability is best suited for assessing compaction since it is particularly sensitive to changes in the coarse pore fraction; sorptivity is sensitive to surface phenomena such as the effects of curing, while chloride conductivity provides good characterization of the binder type for marine concretes or concretes exposed to chlorides.

6.5.1.7.2.2 CONCRETE MIX OPTIMIZATION FOR DURABILITY

Durability indexes may be used to optimize materials and construction processes for specific performance criteria. At the design stage, the influence of a range of factors may be evaluated for their impact on quality of the concrete, in particular, the cover layer. A cost-effective solution to ensuring durability may in this way be assessed using a rational testing strategy.

Table 6.4 Initial scheme: suggested ranges for durability classification (in terms of quality of the cover concrete) using index values

Durability class	OPI (log scale)	Sorptivity (mm/√h)	Conductivity (mS/cm)
Excellent	>10	<6	<0.75
Good	9.5–10	6–10	0.75–1.50
Poor	9.0–9.5	10–15	1.50–2.50
Very poor	<9.0	>15	>2.50

6.5.1.7.2.3 PERFORMANCE-BASED SPECIFICATIONS

From controlled laboratory studies and site data, a matrix of durability index parameters has been developed to produce a set of acceptance and rejection criteria for performance-based specifications (further details given later). An early scheme with suggested ranges for durability classification of concretes, based on site and laboratory data, is shown in Table 6.4; however, this has since been refined and improved, and is given for illustrative purposes.

6.5.1.7.2.4 PREDICTIONS OF LONG-TERM PERFORMANCE: SERVICE LIFE MODELS

Service life predictions of RC structures are affected by many variables that make precise estimates of durability performance difficult. Since durability index tests are based on transport mechanisms associated with deterioration, it follows that these indexes can be used for durability prediction. The durability indexes have been related empirically to service life prediction models. The indexes are used as input parameters together with other variables such as cover and environmental class, to determine notional design life. Limiting index values can be used in construction specifications to provide the necessary concrete quality for a required life and environment.

Two corrosion initiation models have been developed, for carbonation- and chloride-induced corrosion, deriving from measurements and correlations of short-term durability index values, aggressiveness of the environment, and actual deterioration rates monitored over periods of up to 10 years. The models allow for determining the expected life of a structure based on environmental conditions, cover thickness, and concrete quality. Environmental classification is based on EN 206, while concrete quality is represented by the appropriate durability index parameter. The OPI is used in the carbonation prediction model, and the CCI in the chloride model. The SLMs can also be used to determine the required value of the durability parameter based on predetermined values for cover thickness, environment, and expected design life. Alternatively, if concrete quality is known from the appropriate DI, a corrosion-free life can be estimated for a given environment.

Despite the proven usefulness of the DI approach, caution demands that durability index tests should not be used indiscriminately when making durability predictions. Comparisons between different concretes in particular may be misleading unless long-term effects are considered. Durability predictions should be made only when the relationship between early-age properties and medium- or long-term durability performance has been established. While the existing SLMs attempt to do this, further refinement is needed.

The link between durability index parameters and their use in service life prediction models is important, since this represents an integrated approach, mentioned earlier, that allows for continual improvement and modification as additional data become available. It also introduces a powerful tool by linking material properties directly with expected service life, and allows decisions to be taken at the construction stage based on measured material parameters, for example, whether to implement additional protective measures if the material fails the required DI test values. It also represents a rational approach to ensuring the achievement of concrete quality by being linked to performance-based specifications.

6.5.1.7.3 Implementation

An example of a large-scale implementation is given in Nganga et al. (2013). The SANRAL implemented DI-based performance specifications in a major freeway infrastructure project, involving, *inter alia*, construction of bridges and interchanges. Durability-sensitive structures (precast median barriers and structural elements such as bridge piers, decks, and abutments) were required to meet durability criteria using DI tests and cover depth measurements, in addition to strength criteria.

The limiting values in the specifications were a minimum of 9.70 (log scale) for OPI, a maximum of 10 mm/hour$^{0.5}$ for WSI, and an average of 40 mm for cover depth, with suitable allowance for statistical variance. These limits were obtained from the service life prediction model for carbonation, considering the relevant exposure conditions (no chlorides) and binder type used. Using the SLM, and for a service life of 100 years, an OPI of 9.70 would yield a predicted carbonation depth of 28 mm, giving latitude on the required cover depth of 40 mm.

The specified limiting values and reduced payment criteria are summarized in Table 6.5. For the water sorptivity test, no reduced payment criterion was applied as the test is used only as an internal control to monitor the effectiveness of curing.

DI tests were carried out on cores removed from either test panels or the actual structures (e.g., precast median barriers). Test panels (400 mm wide, 600 mm high, 150 mm thick) were constructed with the same concrete, shutter type, compaction, and curing methods used for the actual structure, placed adjacent to the structure, with cores removed after 28–35 days.

Table 6.5 Limiting values in DI-based performance specifications and reduced payment criteria (SANRAL)

| | Oxygen permeability index | | Concrete cover (40 mm) | |
	OPI (log scale)	Percentage payment	Overall cover (mm)	Percentage payment
Full acceptance	>9.70	100%	≥85% <(100% + 15 mm)	100%
Conditional acceptance[a]	>8.75 ≤9.70	80%	<85% ≥75%	85%
Conditional acceptance[b]	–	–	<75%	70%
Rejection	<8.75	Not applicable	<65%	Not applicable

[a] With reduced payment.
[b] With remedial measures as approved by engineer, and reduced payment.

OPI results are given in Table 6.6 and in Figure 6.16, by way of summary statistics from five subprojects. Projects 1, 2, 4, and 6 had samples from *in situ* structures (panels) while Project 9 was from precast elements.

While mean OPI values for all the projects exceeded the minimum value of 9.70, there was a high proportion that failed to comply (defectives), with as much as 40.1% for Project 1 and 26.4% for Project 6. For Project 9 (precast elements), all values complied with the limit value. Variability (CoV) of OPI ranged from 1.75% to 4.60%. These values can be compared with a previous inter-laboratory exercise where the reproducibility of the OPI test was determined as 1.8%. Site-based results can be expected to have a higher variability; on this basis, Project 6 has an unacceptably high variability. The data show that the OPI test is a useful and sensitive parameter to measure construction quality variation on-site, able to detect different

Table 6.6 Summary statistics of OPI values from SANRAL projects

| | | OPI (log scale) | | | | | Proportion of defectives[c] |
Project ID	n[a]	Mean	Max	Min	S[b]	CoV (%)	%
1	172	9.75	10.41	9.07	0.28	2.84	40.1
2	94	9.91	10.42	9.37	0.22	2.24	13.8
4	116	9.87	10.40	9.39	0.23	2.33	18.1
6	91	10.06	11.10	8.83	0.46	4.60	26.4
9	132	10.25	10.70	9.85	0.18	1.75	0

Source: Nganga, G., Alexander, M. G., and Beushausen, H., 2013, Practical implementation of the durability index performance-based design approach, *Construction and Building Materials*, 45, 251–261.

[a] Number of results.
[b] Standard deviation.
[c] Results that failed to comply with the limit OPI value of 9.70.

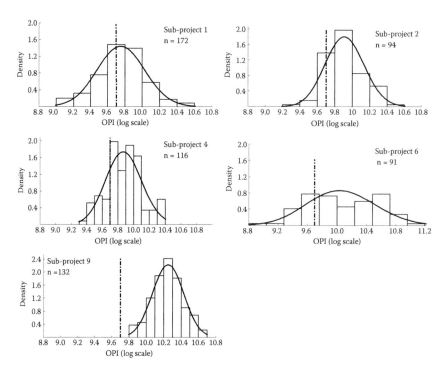

Figure 6.16 Histogram plots of OPI values from five SANRAL projects. (Reproduced from Nganga, G., Alexander, M. G., and Beushausen, H., 2013, Practical implementation of the durability index performance-based design approach, *Construction and Building Materials*, 45, 251–261. With permission from Elsevier.)

practices resulting in very different variability. Where adequate quality control is exercised, such as in the precast operation, full compliance with the specification is achievable.

6.6 CLOSURE

This chapter reviewed durability indicators (or durability indexes), which can encompass physical, mechanical, chemical, and electrochemical properties of concrete. These have been developed largely in response to the need for quantifiable measures of concrete durability, in order to move away from simplistic prescriptive measures that are not adequately verifiable. They can be used in performance-based approaches and in service life prediction models. Further, if properly formulated, they can be used to assess the actual *in situ* quality of a concrete structure, thus giving confidence in the *as-built* performance. Much depends on the existence of reliable and reproducible test methods to measure these indicator properties. Considerable progress has been made in a number of countries in implementing performance-based approaches, which draw to a greater or lesser extent on durability indicators or indexes. Further development can be expected in the future, which will permit the development of full performance-based approaches linked to measures of "durability reliability."

REFERENCES

Alexander, M. G. (ed.), 2016, *Marine Concrete Structures: Design, Durability and Performance*, Woodhead Publishing (Imprint of Elsevier), Cambridge, UK.

Alexander, M. G. and Ballim, Y., 1993, Experiences with durability testing of concrete: A suggested framework incorporating index parameters and results from accelerated durability tests, in *Proc. 3rd Canadian Symp. on Cement and Concrete*, Ottawa, Canada, August 1993, Nat. Res. Council, Ottawa, Canada, pp. 248–263.

Alexander, M. G., Ballim, Y., and Stanish, K., 2008, A framework for use of durability indexes in performance-based design and specifications for reinforced concrete structures, *Materials & Structures*, 41(5), 921–936.

Alexander, M. G. and Beushausen, H., 2009, Performance-based durability testing, design and specification in South Africa: Latest developments, in *Int. Conf. on Excellence in Concrete*, Kingston, London, September 2008, CRC Press, London, UK, pp. 429–434.

Alexander, M. G., Mackechnie, J. R., and Ballim, Y., 1999, *Guide to the Use of Durability Indexes for Achieving Durability in Concrete Structures*, Research Monograph No. 2, Department of Civil Engineering, University of Cape Town, Rondebosch, South Africa, 35pp.

Alexander, M. G., Mackechnie, J. R., and Ballim, Y., 2001, Use of durability indexes to achieve durable cover concrete in reinforced concrete structures, in *Materials Science of Concrete*, J. P. Skalny and S. Mindess (eds.), American Ceramic Society, Westerville, OH, Vol. VI, pp. 483–511.

Andrade, C., 1993, Calculation of chloride diffusion coefficients in concrete from ionic migration measurements, *Cement and Concrete Research*, 23, 724–742.

Andrade, C., 2004, Calculation of initiation and propagation periods of service life of reinforcements by using the electrical resistivity, in *International Symposium: Advances in Concrete through Science and Engineering*, Evanston, Northwestern University, Evanston, IL, p. 8.

Andrade, C., Alonso, C., and Goni, S., 1993, Possibilities for electrical resistivity to universally characterize mass transport processes in concrete, in *Concrete 2000 Economic and Durable Construction through Excellence Volume Two: Infrastructure, Research, New Applications*, R. K. Dhir and M. R. Jones (eds.), E & FN SPON, Scotland, UK, pp. 1639–1652.

Andrade, C., Alonso, C., Gulikers, J., Polder, R., Cigna, R., Vennesland, Ø., Salta, M., Raharinaivo, A., and Elsener, B., 2004, RILEM TC 154-EMC: Electrochemical techniques for measuring metallic corrosion. Recommendations test methods for on-site corrosion rate measurement of steel reinforcement in concrete by means of the polarization resistance method, *Materials & Structures*, 37(273), 623–643.

Andrade, C. and d'Andrea, R., 2010, Electrical resistivity as microstructural parameter for modelling of service life of reinforced concrete structures, in *2nd International Symposium on Service Life Design for Infrastructure, RILEM Proceedings 70, RILEM Publications sarl*, Paris, pp. 379–388.

Andrade, C., Gulikers, J., Polder, R., and Raupach, M., 2003, Half-cell potential measurements—Potential mapping on reinforced concrete structures, *Materials and Structures*, 36, 461–471.

Andrade, C., Rebolledo, N., Castillo, A., Tavares, F., Perez, R., and Baz, M., 2014, Use of resistivity and chloride resistance measurements to assess concrete durability of new Panama Canal, in *RILEM International Workshop on Performance-Based Specification and Control of Concrete Durability, RILEM*, Zagreb, Croatia, p. 9.

ASTM C1202-12, 2012, *Standard Test Method for Electrical Indication of Concrete's Ability to Resist Chloride Ion Penetration*, Vol. 04.02, ASTM International, West Conshohocken, PA.

ASTM C1585-13, 2013, *Standard Test Method for Measurement of Rate of Absorption of Water by Hydraulic-Cement Concretes*, Vol. 04:02, ASTM International, West Conshohocken, PA.

Atkins, P. W. and De Paula, J., 2006, *Physical Chemistry*, 8th edition, Oxford University Press, Oxford.

Ballim, Y., 1993, Curing and the durability of OPC, fly ash and blast-furnace slag concretes, *Materials and Structures*, 26(158), 238–244.

Ballim, Y., 1994, Curing and the durability of concrete, PhD thesis, University of the Witwatersrand, Johannesburg, South Africa.

Baroghel-Bouny, V., 2002, Which toolkit for durability evaluation as regards chloride ingress into concrete? Part II: Development of a performance approach based on durability indicators and monitoring parameters, in *3rd International RILEM Workshop "Testing and Modelling Chloride Ingress into Concrete"*, C. Andrade and J. Kropp (eds.), RILEM, Madrid, Spain, pp. 137–163.

Baroghel-Bouny, V., 2004, Durability indicators: A basic tool for performance-based evaluation and prediction of RC durability, in *International Seminar on Durability and Lifecycle Evaluation of Concrete Structures*, R. Sato, Y. Fujimoto, and T. Dohi (eds.), University of Hiroshima, Higashi-Hiroshima, Japan, pp. 13–22.

Baroghel-Bouny, V., Nguyen, T. Q., Thiery, M., Dangla, P., and Belin, P., 2006, Evaluation and prediction of reinforced concrete durability by means of durability indicators. Part II: Multi-level predictive modelling, in *International RILEM-JCI Seminar on Concrete Durability and Service Life Planning*, K. Kovler (ed.), RILEM Publications SARL, Paris, France, pp. 270–280.

Basheer, P. A. M., Long, A. E., and Montgomery, F. R., 1994, The Autoclam—A new test for permeability, *Concrete*, 28(4), 27–29.

BS 1881: Part 5, 1970, *Test for Determining the Initial Surface Absorption for Concrete*, British Standard Institute, London.

CSA A23.1-14, 2014, *Concrete Materials and Methods of Concrete Construction*, Canadian Standards Association, Toronto, Canada.

CSA A23.2-14, 2014, *Test Methods and Standard Practices for Concrete*, Canadian Standards Association, Toronto, Canada.

CUR Leidraad 1, 2009, Duurzaamheid van constructief beton met betrekking tot chloride-geïnitieerde wapeningscorrosie, CUR Gouda, Durability of structural concrete with respect to chloride induced corrosion, in Dutch, Gouda,

Netherlands: Centre for Civil Engineering Research and Codes (CUR)/ Rijkswaterstaat (RWS).

de Rooij, M. R., Polder, R. B., and van Oosten, H. H., 2007, Validation of durability of cast in situ concrete of the Groene Hart railway tunnel, *HERON*, 52(4), 225–238.

DuraCrete, 1998, Probabilistic performance-based durability design: Modelling of degradation. Document, D. P. No. BE95-1347/R4-5, The Netherlands.

DuraCrete, 2000, DuraCrete, DuraCrete Final Technical Report R17, Document BE95-1347/R17, The European Union—Brite EuRam III, DuraCrete—Probabilistic Performance based Durability Design of Concrete Structures, CUR, Gouda, The Netherlands.

EN 206-1, 2013, *Concrete—Part 1: Specification, Performance, Production and Conformity*, British Standards Institution, London.

Ferreira, R. M., 2004, Probability-based durability design of concrete structure in marine environment, PhD thesis, University of Minho, Guimarães, Portugal.

Ferreira, R. M., Gjørv, O. E., and Jalali, S., 2004, Software for probability-based durability design of concrete structures, in *Proceedings V. 1, Fourth International Conference on Concrete Under Severe Conditions—Environment and Loading*, B. H. Oh, K. Sakai, O. E. Gjørv, and N. Banthia (eds.), Seoul National University and Korea Concrete Institute, Seoul, South Korea, pp. 1015–1024.

fib Model Code for concrete structures, 2010, Ernst & Sohn, Oxford.

Gjørv, O. E., 2003, Durability of concrete structures and performance-based quality control, in *Proceedings, International Conference on Performance of Construction Materials in the New Millennium*, A. S. El-Dieb, M. M. R. Taha, and S. L. Lissel (eds.), Shams University, Cairo.

Gjørv, O. E., 2009, *Durability Design of Concrete Structures in Severe Environments*, Taylor & Francis, London.

Gjørv, O. E., 2010, Durability design and quality assurance of concrete infrastructure—Performance-based programs boost service life, *Concrete International*, 32(9), 29–36.

Gjørv, O. E., 2014, *Durability Design of Concrete Structures in Severe Environments*, 2nd edition, Taylor & Francis, CRC Press, London. Also published in Chinese by Press of China Building Materials Industry, Beijing (2015), and in Portuguese by Oficina de Textos, Sao Paulo (2015).

Gjørv, O. E., 2015, Quality control for concrete durability—A case study provides comparisons of work performed under performance and prescriptive specifications, *Concrete International*, 37(11), 52–57.

Gouws, S. M., Alexander, M. G., and Maritz, G., 2001, Use of durability index tests for the assessment and control of concrete quality on site, *Concrete Beton*, 98, 5–16.

Hooton, R. D. and Bickley, J. A., 2012, Prescriptive versus performance approaches for durability design—The end of innocence, *Materials and Corrosion*, 63(12), 1097–1101.

Hooton, R. D. and Bickley, J. A., 2014, Design for durability: The key to improving concrete sustainability, *Construction and Building Materials*, 67, 422–430.

Hooton, R. D., Hover, K., and Bickley, J. A., 2005, Performance standards and specifications for concrete: Recent Canadian developments, *The Indian Concrete Journal*, 79(12), 31–37.

Imamoto, K., Tanaka, A., and Kanematsu, M., 2012, Non-destructive assessment of concrete durability of the National Museum of Western Art in Japan, Paper 180, Microdurability 2012, Amsterdam, April 11–13, 2012.

ISO 13823-1, 2008, *General Principles on the Design of Structures for Durability*, International Organization for Standardization, Geneva, Switzerland.

Jacobs, F. and Hunkeler, F., 2006, Non-destructive testing of the concrete cover— Evaluation of permeability test data, in *International RILEM Workshop on Performance Based Evaluation and Indicators for Concrete Durability*, V. Baroghel-Bouny, C. Andrade, R. T. Torrent, and K. Scrivener (eds.), RILEM Publications SARL, Madrid, Spain, p. 8.

Jacobs, F., Leemann, A., Denarié, E., and Teruzzi, T., 2009, *SIA 262/1. Recommendation for the quality control of concrete with air permeability measurements*, VSS report, Zurich, Switzerland, 22pp.

Japan Concrete Institute, 2014, Performance evaluation guidelines of existing concrete structures, 2014, 459pp (in Japanese).

Kelham, S., 1988, A water absorption test for concrete, *Magazine of Concrete Research*, 40(143), 106–110.

LIFE-365, 2005, *ACI-Committee-365, Service Life Prediction Model, Computer Program for Predicting the Service Life and Life-Cycle Costs of Reinforced Concrete Exposed to Chlorides*, American Concrete Institute, Farmington Hills, MI.

Lobo, C., Lemay, L., and Obla, K., 2005, Performance-based specifications for concrete, *The Indian Concrete Journal*, 79(12), 13–17.

Millard, S. G., Ghassemi, M. H., Bugey, J., and Jafar, M. I., 1990, Assessing the electrical resistivity of concrete structures for corrosion durability studies, in *Corrosion of Reinforcement in Concrete*, C. L. Page, K. W. J. Treadway, and P. B. Bamforth (eds.), Elsevier, New York, NY, pp. 303–313.

Muigai, R. N., Moyo, P., and Alexander, M. G., 2012, Durability design of reinforced concrete structures: A comparison of the use of durability indexes in the deemed-to-satisfy approach and the full-probabilistic approach, *Materials & Structures*, 45, 1233–1244.

Muigai, R. N., Moyo, P., Alexander, M. G., and Beushausen, H. D., 2009, Application of durability indexes in probabilistic modelling of chloride ingress in RC members, in *Proceedings of the 2nd International RILEM Workshop on Concrete Durability and Service Life Planning ConcreteLife'09*, Haifa, 7–9 September 2009, RILEM Publications SARL, Bagneux, France, pp. 408–415.

Neville, A. M., 1987, Why we have concrete durability problems, in *ACI SP-100, Katherine and Bryant Mather International Conference on Concrete Durability*, ACI, Detroit, MI, 1987, pp. 21–48.

Nganga, G., Alexander, M. G., and Beushausen, H., 2013, Practical implementation of the durability index performance-based design approach, *Construction and Building Materials*, 45, 251–261.

NT Build 443, 1995, *Concrete, Hardened: Accelerated Chloride Penetration*, Nordtest, Finland, 5pp.

NT Build 492, 1999, *Concrete, Mortar and Cement Based Repair Materials: Chloride Migration Coefficient from Non-Steady State Migration Experiments*, Nordtest, Finland, 8pp.

Papadakis, V. G., Vayenas, C. G., and Fardis, M. N., 1991, Fundamental modelling and experimental investigation of concrete carbonation, *ACI Materials Journal*, 88(4), 363–373.

PIANC Norway/NAHE, 2009, Durable concrete structures—Part 1: Recommended specifications for new concrete harbor structures, in *Part 2: Practical Guidelines for Durability Design and Concrete Quality Assurance*, 3rd edition, Norwegian Association for Harbor Engineers (NAHE), TEKNA, Oslo, Norway (in Norwegian).

Polder, R. B. and de Rooij, M. R., 2005, Durability of marine concrete structures—Field investigations and modelling, *HERON*, 50(3), 133–143.

Polder, R. B., de Rooij, M. R., and van Breugel, K., 2006, Validation of models for service life prediction—Experiences from the practice, in *International RILEM-JCI Seminar on Concrete Durability and Service Life Planning ConcreteLife'06*, K. Kovler (ed.), RILEM Publications SARL, Ein-Bokek, Dead Sea, Israel, pp. 292–301.

SANS 3001-CO3-1, 2015, *Civil Engineering Test Methods Part CO3-1: Concrete Durability Index Testing—Preparation of Test Specimens*, South African Bureau of Standards, Pretoria, South Africa.

SANS 3001-CO3-2, 2015, *Civil Engineering Test Methods Part CO3-2: Concrete Durability Index Testing—Oxygen Permeability Test*, South African Bureau of Standards, Pretoria, South Africa.

SANS 3001-CO3-3, 2015, *Civil Engineering Test Methods Part CO3-3: Concrete Durability Index Testing—Chloride Conductivity Test*, South African Bureau of Standards, Pretoria, South Africa.

Sengul, O. and Gjørv, O. E., 2008, Electrical resistivity measurements for quality control during concrete construction, *ACI Materials Journal*, 105, 541–547.

Somerville, G., 1997, Engineering design and service life: A framework for the future, in *Prediction of Concrete Durability: Proceedings of STATS 21st Anniversary Conference*, J. A. N. Glanville (ed.), E & FN SPON, London, pp. 58–76.

Stanish, K., Alexander, M. G., and Ballim, Y., 2006, Assessing the repeatability and reproducibility values of South African durability index tests, *Journal of the South African Institution of Civil Engineering*, 48(2), 10–17.

Streicher, P. E. and Alexander, M. G., 1995, A chloride conduction test for concrete, *Cement and Concrete Research*, 25(6), 1284–1294.

Swiss Standard SN 505 262/1, 2013, Concrete Construction—Complementary Specifications, Schweizer Norm, 1 August 2013, 52pp (in German and French).

Torrent, R. J., 1992, A two-chamber vacuum cell for measuring the coefficient of permeability to air of the concrete cover on site, *Materials & Structures*, 25(6), 358–365.

Torrent, R., 2015, Exp-Ref: A simple, realistic and robust method to assess service life of reinforced concrete structures, in Concrete 2015, Melbourne, Australia, August 30 to September 2.

Torrent, R., Armaghani, J., and Taibi, Y., 2013, Evaluation of Port of Miami Tunnel Segments: Carbonation and service life assessment made using on-site air permeability tests, *Concrete International*, May, 39–46.

Torrent, R., Denarié, E., Jacobs, F., Leemann, A., and Teruzzi, T., 2012, Specification and site control of the permeability of the concrete cover: The Swiss approach, *Materials and Corrosion*, 63(12), 1127–1133.

Torrent, R. and Imamoto, K., 2015, Site testing of air-permeability as indicator for carbonation rate in old structures, in *Intern. Confer. Regeneration and Conservation of Concr. Struct.*, Nagasaki, Japan.

Torrent, R. and Jacobs, F., 2014, Swiss Standards 2013: World's most advanced durability performance specifications, in *3rd Russian Intern. Confer. on Concr. and Ferrocement*, Moscow, Russia.

Tramex, 2014, Concrete Encounter CME4, available at: http://www.tramex.ie/Our_Products_Details.aspx?ID=19, retrieved March 15, 2014.

University of Cape Town (UCT), Department of Civil Engineering, 2009, *Durability Index Testing Procedure Manual*, 30pp.

van der Wegen, G., Polder, R. B., and van Breugel, K., 2012, Guideline for service life design of structural concrete—A performance based approach with regard to chloride induced corrosion, *HERON*, 57(3), 153–166.

Wang, Y. F., Dong, G. H., Deng, F., and Fan, Z. H., 2014a, Application research of the efficient detection for permeability of the large marine concrete structures, *Applied Mechanics and Materials*, 525, 512–517.

Wang, P. P., Dong, G. H., Li, K. F., and Torrent, R., 2014b, *Durability Quality Control for Immerged Tube Concrete through Air Permeability and Electrical Resistance Measurement*, SLD'14, Service Life Design for Infrastructure, Zhuhai, China.

FURTHER READING

Alexander, M. G., Ballim, Y., and Santhanam, M., 2005, Performance specifications for concrete using the durability index approach, *The Indian Concrete Journal*, 79(12), 41–46.

Alexander, M. G. and Nganga, G., 2013, Reinforced concrete durability: Some recent developments in performance-based approaches, in *Keynote Paper, ISCC 2013 (International Seminar on Cement and Concrete)*, Nanjing, China, September 2013, 13pp.

Alexander, M. G. and Santhanam, M., 2013, Achieving durability in reinforced concrete structures: Durability indices, durability design and performance-based specifications, in *Keynote Paper at International Conferences on Advances in Building Sciences & Rehabilitation and Restoration of Structures*, IIT Madras, Chennai, India, Feb 2013, 21pp.

Alexander, M. G., Santhanam, M., and Ballim, Y., 2011, Durability design and specification for concrete structures—The way forward, *International Journal of Advances in Engineering Sciences and Applied Mathematics*, 2(3), 95–105.

Alexander, M. G., Stanish, K., and Ballim, Y., 2006, Performance-based durability design and specification: Overview of the South African approach, in *International RILEM Seminar on Performance Indicators*, Madrid, Spain.

Beushausen, H. and Alexander, M. G., 2009, Application of durability indicators for quality control of concrete members—A practical example, in *Proceedings of the International RILEM TC-211 PAE Final Conference*, Toulouse, France, June 2009, RILEM Publications SARL, Bagneux, France, pp. 548–555.

Nganga, G., Alexander, M. G., and Beushausen, H., 2011, Performance-based durability design for RC structures, *3R's Journal*, 2(3), 284–290.

Chapter 7

Durability testing

Transport properties

7.1 INTRODUCTION

Durability testing has been for many decades an important element in concrete research and specifications. Concrete structures are expected to have a service life of at least several decades and they are very difficult and costly to repair if degradation occurs. In view of the many durability challenges that modern concrete technology is facing, it is now well realized that there is a need to design and specify for durability, and it is also necessary to evaluate the long-term performance of new materials and technologies before they are accepted for practice. Test methods for these purposes are required and they should also support research, standardization, design, and quality control. The advent of the specification approach calls for new types of testing to enable the application of new performance codes.

The need for durability testing encompasses many aspects and purposes, and thus numerous test methods have been developed and used. They may be classified into two types, one assessing the rate and extent of the penetration of the degradation agents (i.e., the transport properties of the concrete with respect to specific agents) and the other identifying and quantifying the mechanisms and distress induced by the degrading agents. The first type of tests will be reviewed in this chapter and the second type in Chapter 8.

The concepts upon which the tests are developed are quite varied. Generally, in the modern tests, the intent is to base them on sound physical and chemical principles. Such parameters, which are of quantitative significance in nature, could be used for modeling both to predict degradation processes and quantitatively to assess changes in performance over time in aggressive environments, to predict and design for service life. However, this is not always possible, especially when the intention is for a test which is easy to use for routine purposes. Such tests provide outputs which could be applied for comparative purposes or for prescription-based standards.

Attempts have been made to introduce accelerating effects, such as increased temperature and higher concentrations of degradation agents. The intention when such means are taken is to provide test methods which can yield significant results in a relatively short time. Investigations have been carried out to correlate the results of the accelerated tests with natural exposure.

Overviews of these tests are presented in many references, and those that have been particularly used here are ACI (2000, 2008), Bickley et al. (2006), DuraCrete (2000), DARTS (2004a,b), Lobo (2007), Carino (1999, 2004), Wong (2007), Bentur et al. (1997), Torrent and Luco (2007), and RILEM TC116 (1999).

7.2 PENETRATION TESTS

The penetration tests can be classified according to the type of fluid involved (gas, water, aqueous solution) as well as the nature of the penetrating medium, whether chemically reactive with the cement matrix (e.g., chlorides, CO_2) or whether largely inert (e.g., H_2O, N_2, O_2).

7.2.1 Nonreactive penetrating fluids

When nonreactive fluids are used, the results of the penetration test can be interpreted on the basis of the pore structure by considering models of physical interactions between the penetrating substances and the pores. However, these interactions are largely dependent on the degree of saturation of the pores, which can evoke different types of penetration mechanisms and lead to outputs, which depend on the moisture content of the pores. For example, in partially dried pore structures, the mechanism of H_2O penetration is to a large extent of the convection-capillary type and the extent of penetration and the measured results would be very sensitive to the moisture content of the tested material. In the case of penetration of gas, the mechanism in a saturated concrete would be mainly of the diffusion type and would be vastly different in terms of modeling and penetration rates from unsaturated partially dry concrete. Such influences require special precautions in the testing procedures and the interpretation of the results. In the case of laboratory tests, these sensitivities to the moisture state require conditioning of the samples in predetermined moisture conditions.

Site tests must be accompanied by determination of the moisture content of the tested concrete. The determined moisture content should be part of the report and the interpretation of the results. An overview of the effect of conditioning and recommendations for lab preconditioning are provided in RILEM TC116 (1999). In practice, this sensitivity to moisture content is a considerable limitation for site studies. A potential alternative solution is to use saturated concretes, or to wet the concrete surface on-site. However, in such instances, the penetration of water will not take place and diffusion of gas is far too slow to provide meaningful results in a reasonable time. An alternative that has been gaining some acceptance in recent years is based on testing the electrical conductivity in the saturated concrete, which is a quick and simple test to carry out, and the results might be used for assessing the pore structure density. Data of this kind are relevant for the estimation of the resistance of the concrete to penetration of degrading agents, and it might be used as an input parameter in corrosion of steel modeling where the processes are electrochemical in nature. Since the electrical resistance is highly sensitive to moisture content, it has also been used as a means to assess the moisture content of the concrete. Knowing this value can serve as a means for assisting in the interpretation of the test results of penetration of fluids, to account for the effect of the moisture content and normalize for it.

It should be noted that the application of site tests of this kind is usually carried out on the concrete surface and as such the values obtained reflect the properties of the concrete cover and not those of the bulk concrete. The difference between the bulk and cover concrete can be quite big, especially in cases where segregation and bleeding have taken place and proper water curing was not practised. All of these emphasize the need for caution in the interpretation of test results and the requirement of expertise and understanding, which go far beyond the simple operation of the testing devices.

7.2.2 Reactive penetrating fluids

There is special interest in the evaluation of specific fluids which evoke degradation processes. Determination of their rate of penetration is an important input for modeling of service life, as well as the assessment of means and technologies to slow down their penetration. Most important are the penetration of aqueous chloride solutions and CO_2 gas, which can lead to depassivation of the steel reinforcement to initiate corrosion. The chloride ions penetrating into the concrete can react with hydration products, mainly the aluminates, and as a result measurements of rates of penetration cannot be simply interpreted in terms of physical processes. More than that, since it is only the free chloride ions in solution that are available for depassivation, estimates that do not distinguish between the total chlorides penetrated and the part of it which has been bound may lead to underestimation of the service life.

The situation is quite different with regards to CO_2 penetration where the impact of penetration is not simply the amount penetrated but rather its interaction with hydration products, to carbonate them. The result of carbonation is the reduction in the pH of the pore solution that is in contact with the pore walls. Therefore, in this case, the actual measurement is not the CO_2 content, but rather the change in the pH of the pore solution as a function of the depth from the concrete surface and the elapsed time.

Simultaneous penetration and reaction processes present particular difficulties in quantification of these effects in terms of penetration parameters having basic physical significance. It is difficult to resolve between the physical process of penetration and the chemical reaction of the penetrating fluid. Therefore, it is quite common to quantify these cases in terms of an "effective penetration parameter" by assuming that the penetration is only driven by a physical effect, and the penetrating parameter calculated is an "apparent" one. The "apparent" penetration parameter quantifies within it the chemical effect and can thus be used more readily for simplified engineering-type modeling.

7.3 PENETRATION TESTS BASED ON TRANSPORT OF WATER

The coefficient of water permeation might be determined based on the D'Arcy law if the concrete specimen is subjected to a water pressure difference and the flow through the concrete is measured, preferably taking the time until steady-state flow is achieved. However, the penetration of water under pressure is an extremely slow process and it could take many months or even years to achieve measurable flow of water through concrete specimens (El-Diab and Hooton, 1995). Such a test represents the transport into the concrete when it is fully saturated, and this is rarely the case in practice. In unsaturated concrete, the convection mechanisms governed by capillarity are the most significant ones, and thus tests based on the evaluation of capillarity are more commonly used to assess the quality of the concrete and provide a basis for its specifications in a variety of schemes such as the durability index approach. Such tests are reviewed in Torrent et al. (2007) and some of them are covered in various standards such as DIN 52617 and ASTM C1585, based on measuring capillary rise in unsaturated concrete.

The laboratory capillary rise test methods are based on immersion of the concrete specimen into a water bath to a depth of few millimeters (~2 mm) and monitoring periodically the weight change or the water front as water is being absorbed into the concrete by the mechanism of capillary rise (Figure 7.1).

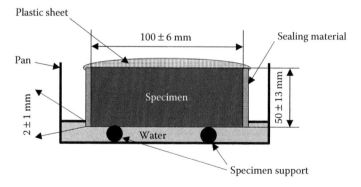

Figure 7.1 Schematic description of the capillary absorption test. (After ASTM 1585.)

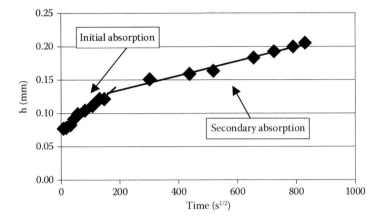

Figure 7.2 Plots obtained in capillary absorption tests in terms of capillary rise versus square root of time. (After ASTM 1585.)

It can be predicted from the capillarity theory that the absorbed water content would be linearly related with the square root of time, and therefore the test results are presented in the form of plots of weight increase or water rise as a function of the square root of time (Figure 7.2).

The test results are quantified in terms of a single parameter, which is the initial slope of the absorption versus square root of time curve. This slope is called the coefficient of capillary absorption.

Since the process reflects the behavior in unsaturated conditions, the test results are sensitive to the moisture content in the concrete (Bentur et al., 2006). Thus, the standards for the laboratory tests provide requirements for the conditioning of the concrete prior to the test and these are different for the various standards, as demonstrated in Table 7.1. Attention should be given not only to the drying itself but also to the conditions that facilitate uniform distribution of the moisture in the concrete cross section.

Test results are very sensitive to the moisture conditions as seen in Table 7.2, which presents a comparison of results obtained with the German

Table 7.1 Requirements for conditioning of concrete specimens prior to testing for capillary absorption

Test method	Conditioning prior to water absorption
ASTM C1585-13 (drying at 50°C/80%RH)	Specimens are placed at 50°C, 80%RH environment for 3 days, then stored in sealed containers for at least 2 weeks to obtain uniform moisture distribution
DIN 52617 (50°C oven drying)	Specimens are oven dried at 50°C until no mass loss

Table 7.2 Capillary water absorption coefficients obtained by the ASTM and DIN test methods as a function of the content of water-repelling admixtures (DPI and DP2), expressed in units of $g/m^2s^{1/2}$

Mix designation	ASTM C1585	DIN 5617
Control	1.33	9.07
5 L/m³ DP2	0.54	4.14
10 L/m³ DP2	0.28	2.05
15 L/m³ DP2	0.26	1.66
5 L/m³ DPI	0.54	3.60
10 L/m³ DPI	0.42	3.15
15 L/m³ DPI	0.41	2.50

Note: Note that ASTM C1585 converts mass data to $mm/s^{0.5}$, which was not used here to keep values consistent with the other test.

Source: Adapted from Bentur, A., Berke, N. S., and Li, L., 2006, Integration of technologies for optimizing durability performance of reinforced concretes, in *Concrete Durability and Service Life Planning (ConcreteLife 06)*, K. Kovler (ed.), *Proceedings RILEM International Conference*, Israel, RILEM Publications PRO 46, Paris, pp. 247–258.

and U.S. test methods for mixes with various contents of water-repelling agents. The values are much higher in the German test in which the drying conditions are harsher. This clearly demonstrates the need to evaluate the results of such tests with caution and consider the moisture conditions in which they were carried out.

Absorption tests using different testing configurations were developed with the view of adopting them for evaluations in the laboratory as well as in the field, as shown for the initial surface absorption test (ISAT) in Figure 7.3, for the laboratory.

The test, which is specified as BS 1881 part 5, is based on measuring the flow along the capillary tube after filling and closing of the water reservoir. Sets of readings are taken at intervals of 10 minutes, 30 minutes, and 1 hour after the first wetting of the surface. There is also an optional 2-hour reading. The test gives sufficiently reliable results when the moisture condition of the concrete is constant; drying of concrete to a constant weight at 105°C is the most effective method. In BS 1881, constant weight is defined when the weight changes are less than 0.1%–0.2% over a 24-hour drying period. Oven drying at 105°C to a constant weight reduced the variability of the results compared to drying at 50°C. A review of the applicability of this test and its repeatability is provided by Basheer et al. (2007) and Dhir et al. (1987).

A similar type of surface absorption testing can be performed using configurations in which there is a small intrusion into the surface of the concrete, such as the Figg test method shown in Figure 7.4 (Figg, 1973). It consists of drilling a small hole ($\varnothing = 10$ mm, depth $= 30$ mm) in the

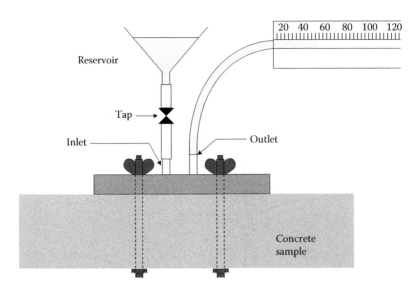

Figure 7.3 Initial surface absorption test scheme for lab testing. (Adapted from BS 1881 Part 5 reviewed in Basheer, M., Goncalez, A. F., and Torrent, R., 2007, Non-destructive methods to measure water transport, in *Non-Destructive Evaluation of the Penetrability and Thickness of the Concrete Cover—State-of-the-Art Report of RILEM Technical Committee 189-NEC*, R. Torrent and L. F. Luco (eds.), RILEM Publications SARL, Bagneux, France.)

concrete. The hole is made airtight by inserting a special silicon rubber plug, which leaves a 20-mm-long cylindrical chamber.

A water head of 100 mm is created above the concrete surface and the water is forced into the assembly using a syringe. The syringe is shut off 1 minute after first contact between water and concrete is made. The time for the meniscus in a capillary tube to travel 50 mm is recorded and is taken as a measure of the water absorption of the concrete.

As noted above, the results of the surface absorption tests are very sensitive to the moisture content in the concrete, and unless account for this is taken for the interpretation of the site tests, the values obtained have limited significance. Several attempts have been reported to make corrections of this kind, for example, DeSouza et al. (1997). In practice, however, the determination of the surface moisture on-site and making the adjustments for it by calculation are extremely difficult to perform. Thus, tests of this kind are at best useful for providing a broad indication of the quality of the concrete, and some guidelines of this nature have been suggested for the ISAT and Figg tests, shown in Table 7.3. Tests of a similar nature such as the Autoclam and GWT-4000 have been reviewed by Basheer et al. (2007)

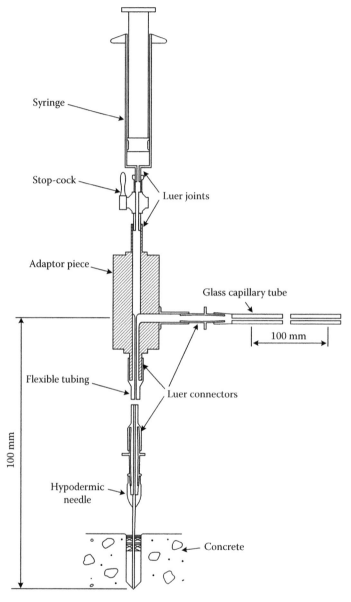

Figure 7.4 Figg surface permeability test configuration which is used for water absorption tests. (Adapted from Figg, J. W., 1973, Methods of measuring the air and water permeability of concrete, *Magazine of Concrete Research*, 25(85), 213–219; Basheer, M., Goncalez, A. F., and Torrent, R., 2007, Non-destructive methods to measure water transport, in *Non-Destructive Evaluation of the Penetrability and Thickness of the Concrete Cover—State-of-the-Art Report of RILEM Technical Committee 189-NEC*, R. Torrent and L. F. Luco (eds.), RILEM Publications SARL, Bagneux, France.)

Table 7.3 Broad classification of the concrete cover quality based on ISAT and Figg poroscope tests

Concrete quality		Bad	Fair/medium	Good/very good
Figg poroscope Absorption time (s)		<20	20–50/50–100	100–500/>500
ISAT value (mL/m²/s) after:	10 min	>0.50	0.25–0.50	<0.25
	30 min	>0.35	0.17–0.35	<0.17
	60 min	>0.20	0.10–0.20	<0.10

Source: Adapted from Concrete Society Working Party, 1987, Permeability testing of site concrete—A review of methods and experience, Concrete society Technical Report No. 31, London, UK; Basheer, M., Goncalez, A. F., and Torrent, R., 2007, Non-destructive methods to measure water transport, in *Non-Destructive Evaluation of the Penetrability and Thickness of the Concrete Cover—State-of-the-Art Report of RILEM Technical Committee 189-NEC*, R. Torrent and L. F. Luco (eds.), RILEM Publications SARL, Bagneux, France.

7.4 PENETRATION TESTS BASED ON TRANSPORT OF NONREACTIVE GASEOUS FLUIDS

The most common gases used for testing for penetration into concrete are air, oxygen, and CO_2. The air and oxygen tests are largely independent of chemical interactions and thus have been developed and used to provide an indication for the pore structure of the concrete with respect to assessment of its quality to act as a barrier to penetration of degrading agents.

Ideally, an apparatus for this purpose should be devised to operate under steady-state conditions of pressure difference and use the D'Arcy law to calculate the coefficient of permeability:

$$\frac{\partial m}{\partial t} = \frac{-k}{g}\left(\frac{\partial P}{\partial z}\right) \tag{7.1}$$

where $\partial m/\partial t$ is the rate of mass flow per unit cross-sectional area, $\partial P/\partial z$ is the pressure gradient in the direction of flow, k is the coefficient of permeability, and g is the acceleration due to gravity.

The most common test configuration for the purpose of determination of the gas coefficient of permeability is the CEMBUREAU test (Kollek, 1989) (Figure 7.5).

Guidance for the testing procedures is provided in RILEM TC116 (1999). They require preconditioning at 50°C until constant weight is achieved. In order to eliminate gradients of humidity within the sample, it is recommended to store the dried specimen in sealed conditions to enable moisture redistribution. The permeability testing is then carried out at three levels of pressure, 0.15, 0.20, and 0.30 MPa, achieving in each steady-state flow. The coefficient of permeability is calculated as the average of the three.

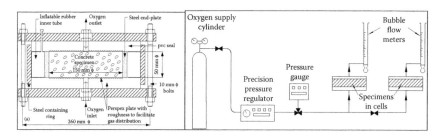

Figure 7.5 Experimental setup and permeation cell of the CEMBUREAU test. (Adapted from Kollek, J. J., 1989, The determination of the permeability of concrete to oxygen by the CEMBUREAU method—A recommendation, *Materials and Structures*, 22, 225–230.)

The time to achieve steady-state flow through the relatively thick concrete specimens may be extremely long. Therefore, testing strategies based on measuring the decay in pressure over a period of time, where the gas on one side of the specimen is pressurized and then the pressure is allowed to decay as gas permeates through the specimen to the other side that is under atmospheric pressure, have been developed. The gradient of pressure over time is measured and the coefficient of permeation is calculated.

A test setup of this kind is shown in Figure 7.6. It was developed within the framework of defining durability indicators based on reliable laboratory testing (Ballim, 1991; Alexander et al., 1999, 2006). The principle of measurement is based on creating a 100 kPa gas pressure in the oxygen vessel cell and thereafter monitoring its decay as gas is permeating through the concrete specimen, to calculate the coefficient of permeability:

$$k = \frac{wVgd}{RA\theta t} \ln \frac{Po}{P} \tag{7.2}$$

where k is the coefficient of permeability (m/s), w is the molecular mass of permeating gas (kg/mol), V is the volume of the pressure cylinder (m³), g is the acceleration due to gravity (m/s²), d is the sample thickness (m), R is the universal gas constant (Nm/K mol), A is the cross-sectional area of specimen (m²), θ is the absolute temperature (K), t is time (s), Po is the pressure at the start of the test (kPa), and P is the pressure at time t (kPa).

Test results demonstrate that a carefully controlled lab test of this kind can serve to differentiate between qualities of concrete such as those affected by curing (Figure 7.7).

Several tests have been developed based on the principle of gas pressure decay with the objective of using them on-site, with a setup placed on the concrete surface. Obviously, the on-site test will apply the natural air as the gaseous medium. An overview of such tests is provided in Torrent et al.

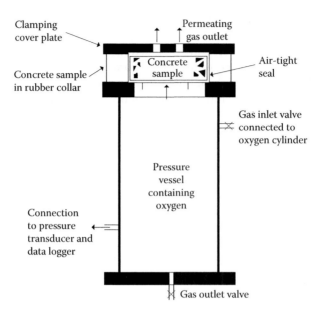

Figure 7.6 Oxygen permeability testing apparatus. (Adapted from Alexander, M. G., Mackenzie, J. R., and Ballim, Y., 1999, Guide to the use of durability indexes for achieving durability in concrete structures, Research Monograph No 2, Department of Civil Engineering, University of Cape Town, 35pp.)

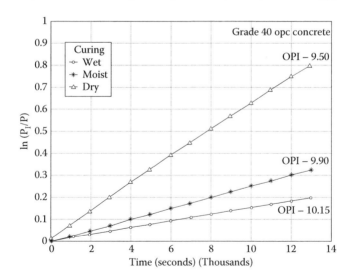

Figure 7.7 Test results for oxygen permeability tests for concrete cured under different conditions. (After Alexander, M. G., Mackenzie, J. R., and Ballim, Y., 1999, Guide to the use of durability indexes for achieving durability in concrete structures, Research Monograph No 2, Department of Civil Engineering, University of Cape Town, 35pp.)

(2007). Some of the more common tests will be described here and the principles are similar to all others.

In the category of these tests are included the Figg test (Figg, 1973) (using an apparatus similar to the water absorption, Figure 7.8) and the Torrent test method, whose principle of operation is presented in Figure 7.9.

This Figg test method configuration (Figure 7.8) is similar to that used for surface absorption of water (Figure 7.4). A hypodermic needle is introduced through the plug into the chamber and a hand vacuum pump produces vacuum inside the chamber. The permeability is assessed in terms of the time in seconds that it takes for the vacuum to fall from −55 to −50 kPa. The advantage of using air rather than water is that readings may be repeated on the same specimen without delay, since the passage of air through the specimen does not change the specimen in any way.

The Torrent air permeability test is also based on creating vacuum and measuring the time required for it to drop down. The test method does not require any drilling or mechanical fastening to the concrete surface (Torrent,

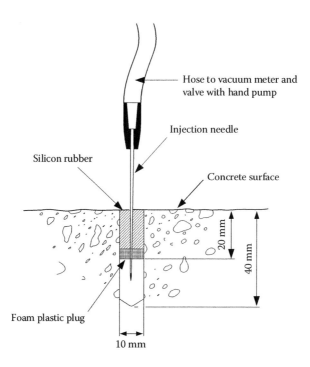

Figure 7.8 Figg surface air permeability test. (After Torrent, R., Basheer, M., and Goncalez, A. F., 2007, Non-destructive methods to measure gas-permeability, in *Non-Destructive Evaluation of the Penetrability and Thickness of the Concrete Cover—State-of-the-Art Report of RILEM Technical Committee 189-NEC*, R. Torrent and L. F. Luco (eds.), RILEM Publications SARL, Bagneux, France.)

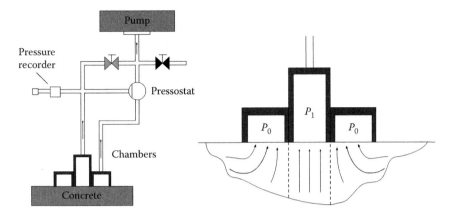

Figure 7.9 Torrent air permeability test. (Adapted from Torrent, R. J., 1992, A two-chamber vacuum cell for measuring the coefficient of permeability to air of the concrete cover on site, *Materials and Structures*, 25(6), 358–365; Torrent, R., Basheer, M., and Goncalez, A. F., 2007, Non-destructive methods to measure gas-permeability, in *Non-Destructive Evaluation of the Penetrability and Thickness of the Concrete Cover—State-of-the-Art Report of RILEM Technical Committee 189-NEC*, R. Torrent and L. F. Luco (eds.), RILEM Publications SARL, Bagneux, France.)

1992). It is based on a two-chamber system, which can be evacuated with a pump (Figure 7.9).

The cell is placed on the concrete surface and vacuum is created in both chambers with the pump. Due to the external atmospheric pressure and the rubber rings, the cell is pressed against the surface and thus both chambers are sealed, making the cell self-supported. After 1 minute, stop-cock 2 (the black one in Figure 7.9) is closed, which isolates the inner chamber system. From this moment on, the pressure in the inner chamber starts to increase, as air is drawn from the underlying concrete. The rate of pressure rise, which is directly related to the permeability of the concrete, is recorded.

Meanwhile, the vacuum pump continues to operate on the outer chamber, through the pressure regulator; the latter ensures that the pressure in the outer chamber is kept always equal to the pressure in the inner chamber. Thus, the outer chamber acts as a "guard-ring," helping to create a controlled air flow into the inner chamber, expected to follow the pattern sketched in Figure 7.9.

Air permeability, kT, is calculated according to the following equation (Torrent, 1992; Torrent and Frenzer, 1995):

$$kT = \left(\frac{V_c}{A}\right)^2 \frac{\mu}{2\varepsilon p_a} \left(\frac{\ln(((p_a + p)/p_a - p)((p_a - p_0)/p_a + p_0))}{\sqrt{t} - \sqrt{t_0}}\right)^2 \tag{7.3}$$

where kT is the air permeability (m^2), V_c is the volume of the test chamber (m^3), A is the cross-sectional area of the test chamber (m^2), μ is the dynamic viscosity of air (N_s/m^2), ε is the porosity of concrete, assumed constant (0.15), p_a is the air pressure (N/m^2), p_o is the air pressure at time t (start of the measurement, after evacuation) in the test chamber (N/m^2), and p is the pressure at time t (end of the measurement) in the test chamber (N/m^2).

Evaluation of the performance of these systems as well as others that are similar in nature, such as the Autoclam, Zia-Guth, and TUD, have been reported in several references (Dhir et al., 1987; Torrent et al., 2007).

Recommendations based on test results from systems of this kind have been suggested to quantify the quality of the concrete in terms of several grades, as demonstrated for the test methods of Figg and Torrent in Table 7.4.

It is essential that specimens for this type of test be conditioned to a uniform and known moisture content in order for the data obtained in the test to be relevant. Practically, this can only be achieved in the laboratory when the specimens are carefully conditioned at the same moisture conditions and means are taken to assure that the moisture is uniformly distributed within the concrete without a gradient at the surface. Variations in moisture content can result in several orders of magnitude differences in the gas permeability (RILEM TC 116, 1999; Romer, 2005) (Figure 7.10).

Attempts to evaluate moisture content by means of electrical resistivity testing, to adjust or normalize for it, are not sufficiently effective to provide reasonably accurate values. At best, they can assist in better qualitative classification of the concrete quality, in terms of the permeability values and the electrical resistance, as demonstrated in Figure 7.11 for the Torrent test method.

Thus, the use of these surface tests on-site has its limitations. Yet, when used in the laboratory for specimens that are conditioned properly, they can yield results that correlate well with the conventional reference tests

Table 7.4 Classification of concrete quality in terms of the Figg air permeability and Torrent tests

Concrete quality	Very good	Good	Medium	Fair	Poor
Figg air permeability (s)	>1000	300–1000	100–300	30–100	<30
Coefficient of air permeability, kT (10^{-16} m^2)	<0.01	0.01–0.1	0.1–1	1–10	>10

Source: Compiled from Figg, J. W., 1988, Discussion of the paper R. K. Dhir, P. C. Hewlett, and Y. N. Chan, Near surface characteristics of concrete: Assessment and development of in situ test methods, in *Magazine of Concrete Research* 1987, 39(141) 183–195, *Magazine of Concrete Research*, 40(145), 234–244; Torrent, R., Basheer, M., and Goncalez, A. F., 2007, Non-destructive methods to measure gas-permeability, in *Non-Destructive Evaluation of the Penetrability and Thickness of the Concrete Cover—State-of-the-Art Report of RILEM Technical Committee 189-NEC*, R. Torrent and L. F. Luco (eds.), RILEM Publications SARL, Bagneux, France.

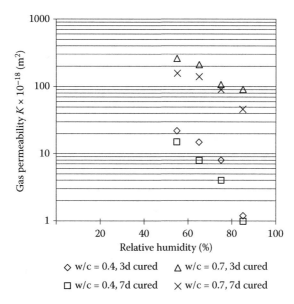

Figure 7.10 Effect of equilibrium moisture concentration on the gas permeability of concrete. (Adapted from RILEM Committee TC 116, 1999, Permeability of concrete as a criterion of its durability, *Materials and Structures*, 32, 174–179.)

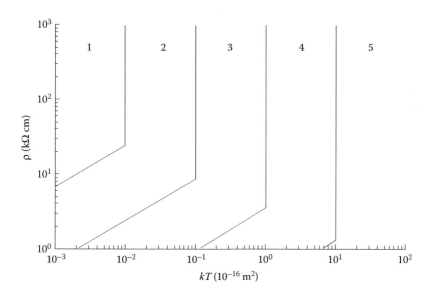

Figure 7.11 Nomogram for the determination of the qualitative category of concrete in terms of electrical resistivity and air permeability. (Based on Torrent, R. J., 1992, A two-chamber vacuum cell for measuring the coefficient of permeability to air of the concrete cover on site, *Materials and Structures*, 25(6), 358–365; courtesy of Proceq SA.)

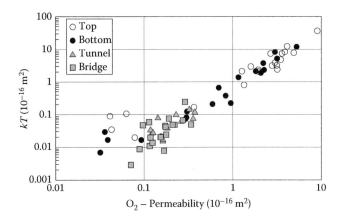

Figure 7.12 Correlation between the air permeability coefficient, *kT*, measured in the Torrent test with the oxygen permeability determined with the CEMBUREAU test. (Adapted from Torrent, R., Basheer, M., and Goncalez, A. F., 2007, Non-destructive methods to measure gas-permeability, in *Non-Destructive Evaluation of the Penetrability and Thickness of the Concrete Cover—State-of-the-Art Report of RILEM Technical Committee 189-NEC*, R. Torrent and L. F. Luco (eds.), RILEM Publications SARL, Bagneux, France; Based on the data of Torrent, R. and Ebensperger, L., 1993, Studie uber Methoden zur messing und Beurteilung der Kennwerte des Uberdeckungsbetons auf der Baustelle—Teil 1 Office Federal de Routes, Suisse, Zurich, January, 119pp; Torrent, R. and Frenzer, G., 1995, Methoden zur messing und Beurteilung der Kennwerte des Uberdeckungsbetons auf der Baustelle—Teil 2 Office Federal de Routes, Suisse, Rapport No. 516, Zurich Oktober, 106pp.)

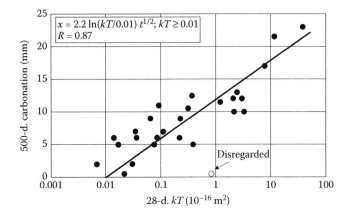

Figure 7.13 Correlations between the air permeability coefficient determined by the Torrent test method and carbonation depth. (Adapted from Torrent, R., Basheer, M., and Goncalez, A. F., 2007, Non-destructive methods to measure gas-permeability, in *Non-Destructive Evaluation of the Penetrability and Thickness of the Concrete Cover—State-of-the-Art Report of RILEM Technical Committee 189-NEC*, R. Torrent and L. F. Luco (eds.), RILEM Publications SARL, Bagneux, France.)

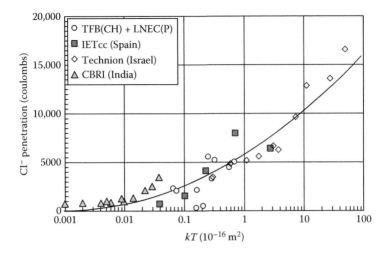

Figure 7.14 Correlation between the air permeability coefficient determined by the Torrent test and chloride penetration by the coulomb tests. (Adapted from Torrent, R., Basheer, M., and Goncalez, A. F., 2007, Non-destructive methods to measure gas-permeability, in *Non-Destructive Evaluation of the Penetrability and Thickness of the Concrete Cover—State-of-the-Art Report of RILEM Technical Committee 189-NEC*, R. Torrent and L. F. Luco (eds.), RILEM Publications SARL, Bagneux, France.)

in which gas is transported through the specimen, as seen for example in Figure 7.12.

These tests might be applied as laboratory tests using equipment, which is more user-friendly than the conventional tests, providing data that are useful for assessing durability parameters, such as carbonation (Figure 7.13) and chloride penetration (Figure 7.14).

In view of these qualities, such tests have been applied to evaluate different technologies for enhancing durability through surface treatments, as shown in Figure 7.15.

7.5 CHLORIDE PENETRATION

The penetration of chlorides into the concrete can take place by a variety of mechanisms, the most important being convection and diffusion. Convection is usually the mechanism by which chloride concentrations are built up on the concrete surface, which is usually in an unsaturated state. The diffusion mechanism largely controls the penetration into the concrete, where it can be more readily considered to be in a saturated state. Chloride ingress is therefore usually modeled on the basis of diffusion mechanisms and the diffusion coefficient needs to be determined by an appropriate test method.

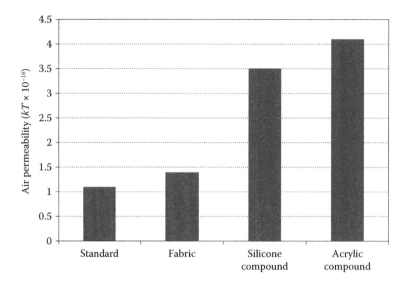

Figure 7.15 Test results of Torrent air permeability demonstrating the effect of curing on concretes that have been oven dried in the lab prior to testing. (After Wasserman, R. and Bentur, A., 2013, Efficiency of curing technologies: Strength and durability, *Materials and Structures*, 46, 1833–1842.)

Ideally, the test to be conducted should create a constant concentration gradient across a concrete specimen and determine the passage of chloride as a function of time. The duration of the test should be such that steady-state conditions be achieved. Achieving steady-state conditions within a reasonable time might only be possible in thin samples of cement paste. However, such samples do not adequately represent concrete specimens, which need to be at least 50 mm thick to provide a representative volume of material in which the aggregates are of the order of 20 mm in size.

In specimens of this size, steady state is not achieved even after many months of test. Therefore, as an alternative, the diffusion coefficient is calculated based on curve fitting of the chloride penetration profiles to the diffusion model equations. The test is carried out by immersion in a chloride solution specimens, which are sealed on all faces except one (ASTM C1556 "Chloride Bulk Diffusion Test," NT BUILD 443 "Accelerated Chloride Penetration") or by ponding on one face of the concrete (AASHTO T259 "Chloride Ponding Test"). It is recommended that the concretes be cured for 14 days under water and thereafter in the laboratory air up to 28 days. After an immersion period of at least 35 days, the profile penetration curve is obtained by grinding off the material in layers parallel to the exposed surface and determining the acid-soluble chloride content in each layer. The standard provides guidance to the thickness of the layers as a function

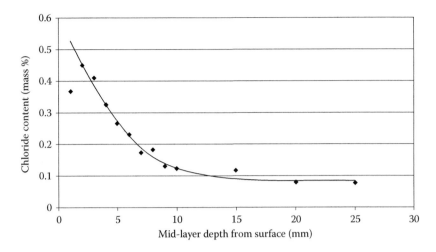

Figure 7.16 Example of regression analysis. (After ASTM C1556.)

of w/c ratio, at intervals of about 1–3 mm for the low w/c ratios (0.40 and lower) to about 2–5 mm for medium and high w/c ratios (0.50–0.70).

The chloride penetration profile obtained (Figure 7.16) is analyzed using regression analysis to fit the chloride diffusion equation to the curve.

The curve fitting is carried out by means of a nonlinear regression least square analysis, by minimizing the sum given in Equation 7.4. Usually, it is recommended to omit the point from the exposed surface layer. The parameters obtained from this analysis are the chloride concentration at the external concrete layer and the diffusion coefficient. It should be emphasized that this coefficient is the non-steady-state diffusion coefficient.

$$S = \sum_{n=2}^{N} \Delta C^2(n) = \sum_{n=2}^{N} (C_m(n) - C_c(n))^2 \tag{7.4}$$

where S is the sum of squares to be minimized (mass %)^2, N is the number of layers ground off, $\Delta C(n)$ is the difference between the measured and the calculated chloride concentration of the nth layer (mass %), $C_m(n)$ is the measured chloride concentration of the nth layer (mass %), and $C_c(n)$ is the calculated chloride concentration in the middle of the nth layer (mass %).

Tests of this kind are mainly of value for research purposes. The limitation of these tests is that they are not entirely representative of diffusion penetration since convection action may take place, especially in the unsaturated AASHTO ponding test.

In order to accelerate such tests, schemes including electrical monitoring have been developed, such as the one shown schematically in Figure 7.17, the ASTM G109 test "Method for Determining Effects of Chemical

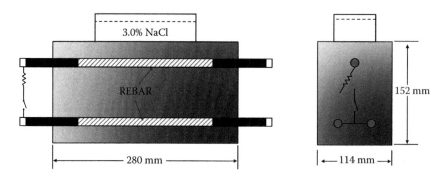

Figure 7.17 Test setup meeting the requirements in ASTM G109 test "Method for Determining Effects of Chemical Admixtures on Corrosion of Embedded Steel Reinforcement in Concrete Exposed to Chloride Environments." (Adapted from Bentur, A., Diamond, S., and Berke, N., 1997, *Steel Corrosion in Concrete*, E&FN SPON/Chapman and Hall, London, UK.)

Admixtures on Corrosion of Embedded Steel Reinforcement in Concrete Exposed to Chloride Environments."

Cyclic ponding is applied on the surface, and electrical measurements of the potential difference between the upper and the lower bars are continuously detected to determine the time at which depassivation occurs at the upper bar (i.e., time of drastic change in the potential to values characteristics of depassivation). The depassivation time can be used for comparative purposes to determine the quality of concrete cover with respect to penetration of chlorides to build up a critical concentration at the level of the steel. This type of test is also lengthy in its duration and serves mainly for research and evaluation of new formulations of concretes or assessment of the influence of cracking on their surface (Figure 7.18).

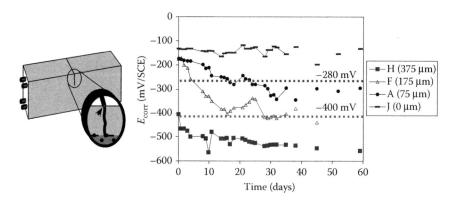

Figure 7.18 Assessment of the influence of cracking on penetration of chlorides using the ASTM G109 test. (After Berke, N. S., 2003, personal communication.)

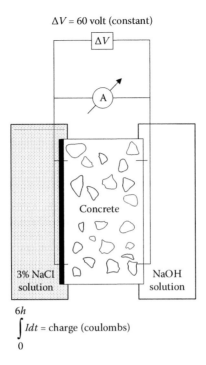

$\Delta V = 60$ volt (constant)

$\boxed{\Delta V}$

Ⓐ

Concrete

3% NaCl
solution

NaOH
solution

$6h$
$$\int_{0} Idt = \text{charge (coulombs)}$$

Figure 7.19 Description of the setup meeting the requirements in ASTM C1202 test.

Alternative test methods, which could be carried out in a reasonable length of time, are based on accelerating the migration of chlorides by means of an electrical potential difference, as seen schematically in Figure 7.19 from the ASTM C1202 test "Test Method for Electrical Indication of Concrete's Ability to Resist Chloride Ion Penetration."

The cell on one side is filled with a 3.0% sodium chloride (NaCl) solution and connected to the negative terminal of the power supply; the cell on the other side contains a 0.3 N sodium hydroxide (NaOH) solution and connected to the positive terminal of the power supply. Chloride ions migrate through the concrete under the influence of the concentration gradient induced by the solutions in the two cells and the electrical potential difference, which is held constant throughout the test at 60 V. The concretes prior to testing are treated to achieve saturation, so the whole test is conducted under saturation conditions. The electrical current is continuously monitored, giving outputs of current versus time such as those presented in Figure 7.20.

The total charge passing in 6 hours (i.e., integration of the current–time curves up to 6 hours, in coulomb units) is used as an indicator of the resistance of the concrete to the passage of chloride ions. Also, electrical resistivity at the onset of the test can be calculated from the initial current and

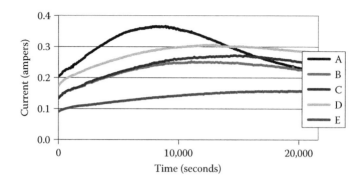

Figure 7.20 Curves of current versus time obtained in the ASTM C1202 test of concrete specimens which have been cured under different water curing regimes, where A is continuous water curing and E being 1 day of water curing. (After Baum and Bentur, A., 2007, internal information, unpublished data.)

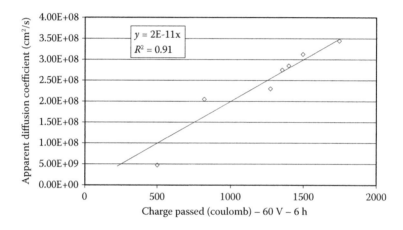

Figure 7.21 Relations between the coulomb value in the ASTM test and chloride diffusion coefficient. (After Medeiros, M. H. F. and Helene, P., 2009, Surface treatment of reinforced concrete in marine environment: Influence on chloride diffusion coefficient and capillary water absorption, *Construction and Building Materials*, 23, 1476–1484.)

the 60 V potential. Significant correlations between the coulomb value and the diffusion coefficient have been reported (Figure 7.21).

A test of a similar nature was developed and used in Europe (Tang and Nilsson, 1992; DuraCrete, 2000; DARTS, 2004a,b) and has become a standard: Nordtest Method, NT Build 492, which is based on the concepts laid down by Tang and Nilsson (1992). It is shown schematically in Figure 7.22. The accelerating principles are the same as for the ASTM C1202 test, and the test setup is slanted to provide means for the escape of gas. However, the mode of operation is different from the ASTM test,

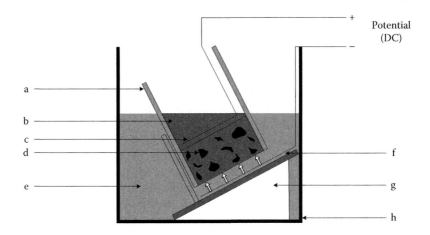

Figure 7.22 Schematic description of the rapid chloride migration test, after the Nordest test. (a) Silicon rubber sleeve; (b) anolyte solution; (c) anode, stainless steel mesh or plate with holes; (d) concrete specimen; (e) catholyte solution; (f) cathode, stainless steel plate about 0.5 mm thick; (g) plastic support under an angle of 32°; and (h) catholyte reservoir, plastic box.

with regards to both the potential–time and concentration regime and the interpretation of the test results.

The potential–time regime is set at a lower potential than the ASTM test and is dependent on the current intensity. At the onset of testing, the potential difference is set at 30 V and the duration of the test is dependent upon the current, with the duration being longer for smaller initial current. This provides means for operating at a lower potential than the ASTM test and thus reduces the influence of heating during the test. The concentration of the catholyte solution is 10% NaCl by mass (100 g NaCl in 900 g water, about 2 N) and the anolyte solution is 0.3 N (approximately 12 g NaOH in 1 L water).

The interpretation of the test results is in terms of the significant physical parameter of the non-steady-state diffusion coefficient. The interpretation is based on the concepts outlined by Tang and Nilsson (1992), who analyzed the system in terms of the Nernst–Planck equation which considers the flux of ionic transport as a function of the concentration gradient and the electrical field:

$$J = -D\frac{dc}{dx} - \frac{zF}{RT}D \cdot c\frac{dE}{dx} \tag{7.5}$$

where D is the diffusion coefficient, Z is the electrical charge, F is the Faraday constant (9.65×10^4 coulomb/mol), R is the gas constant (8.31 J/mol/°K), T is the absolute temperature (°K), E is the electrical voltage (volts), c is the ionic concentration of species in the pore fluid (mol/m³), and x is the distance from the surface (m).

Tang and Nilsson (1992) developed a procedure by which the non-steady-state diffusion coefficient can be calculated from this relation when considering the test results, without the need to determine the Cl⁻ ion profile that occurs during the test, and by relying only on the depth of chloride penetration. The latter can be determined by colorimetric means applied on the split surface of the specimen after the test. Otsuki et al. (1992) demonstrated that the best colorimetric test would be the use of 0.1 N silver nitrate ($AgNO_3$) solution and this was adopted in the NT BUILD 492 standard; at the end of the test, the specimen is split open and the depth of chloride penetration is determined by means of the indicator solution. This solution induces a color change when the chloride concentration is 0.07 mole/L (Otsuki et al., 1992). It reacts with chloride ions to form a white layer, while the zones not penetrated by chlorides are turned brown.

On the basis of these studies, the standard provides the following equation for calculating the non-steady-state chloride diffusion coefficient:

$$D_{RCM} = \frac{RTL}{zFU} \frac{x_d - \alpha \sqrt{x_d}}{t} \tag{7.6}$$

with

$$\alpha = 2\sqrt{\frac{RTL}{zFU}}\, \mathrm{erf}^{-1}\left(1 - \frac{2c_d}{c_o}\right)$$

where D_{RCM} is the rapid chloride migration coefficient (m²/s), z is the absolute value of ion valence (for chloride $z = 1$), F is the Faraday constant ($F = 9.648 \times 10^4$ J·(V mol)⁻¹), U is the absolute value of potential difference (V), R is the gas constant ($R = 8.314$ J (K mol)⁻¹), T is the solution temperature (K), L is the thickness of the specimen (m), x_d is the penetration depth (m), t is the test duration (s), erf⁻¹ is the inverse of the error function, c_d is the chloride concentration for indicator color change ($c_d = 0.07$ mol/L), c_o is the chloride concentration of reagents B in the upstream cell in mol/L, and $\xi = \mathrm{erf}^{-1}[1 - 2(c_d/c_o)]$ (values given in Table 7.5).

The final output is a calculated coefficient D_{RCM} in units of m²/s.

Concern has been raised in the literature that the heating involved in such tests may accelerate the rate of the movement of the ions and result in artificially high values, especially when the measured 6 hours coulomb is above 1500. The lower potential values in the European tests are therefore advantageous.

Table 7.5 Values of ξ

c_o (mol/L)	0.5	1.0	1.5	2.0	2.5	3.0	3.5	4.0	4.5
ξ		0.764	1.044	1.187	1.281	1.351	1.407	1.452	1.491	1.554

The shortcomings due to the heating effect may be avoided if the indicator for chloride migration is chosen as the resistivity measured at the onset of the test. Several reports have shown significant relationships between the resistivity values and the chloride diffusion coefficient (Figure 7.23), and logarithmic correlations of significance have been reported between the coulomb value and resistivity in the ASTM test (Figure 7.24).

Another factor which may affect the results is the pore solution composition of the concrete, since the measured currents in the tests reflect the movement of the chloride ions as well as other ions which may be present in the solution. Thus, when using concrete admixtures which drastically change the pore solution composition, such as calcium nitrite, the coulomb values obtained may be higher than those actually contributed by chloride migration. Therefore, comparison of test results of concretes with and without admixtures should be interpreted with care. The European test that measures the presence of chlorides directly by the $AgNO_3$ solution is therefore advantageous. The AASHTO TP 62 rapid migration test, which is similar to ASTM C1202 in terms of the accelerating voltage conditions, also measures the chloride penetration with the $AgNO_3$ solution.

These test methods, which are relatively simple to carry out within only a few hours, can provide input for the chloride diffusion coefficient to be used in models of service life prediction for durability design. They can also serve as a means for quality control due to their simplicity of execution. Classifications of this kind for the two test methods have been proposed and they are compiled in Table 7.6.

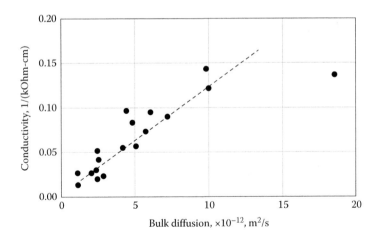

Figure 7.23 Relations between the conductivity (resistivity) and the coefficient of bulk diffusion migration in the European test method. (After Hamilton, R. H. and Boyd, A. J., 2007, Permeability of concrete—Comparison of conductivity and diffusion methods, Research Report, UF Project No. 00026899, Civil and Coastal Engineering, University of Florida.)

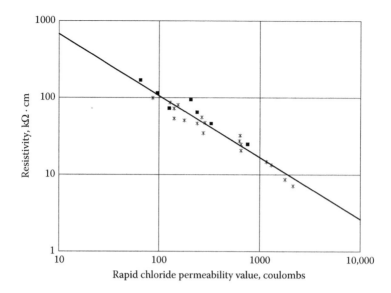

Figure 7.24 Relation between resistivity and coulomb value reported for the ASTM test. (After Bentur, A., Diamond, S., and Berke, N., 1997, *Steel Corrosion in Concrete*, E&FN SPON/Chapman and Hall, London, UK; Adapted from Berke, N. S. and Hicks, M. C., 1992, The life cycle of reinforced concrete decks and marine piles using laboratory diffusion and corrosion data, in *Corrsion Forms and Control for Infrastructure*, V. Chaber, (ed.), ASTM STP 1137, American Society of Testing and Materials, West Conshohocken, PA, pp. 207–231.)

Table 7.6 Classification of chloride penetration resistance of concrete based on different test methods

Chloride penetration classification	Charge (C) American test	Diffusion coefficient (10^{-12} m²/s) European test
High	>4000	100
Moderate	2000–4000	20
Low	1000–2000	10
Very low	100–1000	2
Negligible	<100	1

7.6 PENETRATION OF CO₂ AND CARBONATION

The penetration of CO_2 results in carbonation, which is the consequence of a chemical reaction between the penetrating gas and the hydration products. This reaction changes the composition of the hydration products in contact with the pore solution resulting in a reduced pH of the solution. For these chemical interactions and changes to occur, there is a need for the presence of moisture. Thus, the penetration of CO_2 into drastically dry concrete will not result in any significant carbonation.

The outcome of this process, which is of engineering significance, is the reduction in the pH of the pore solution to values which may facilitate depassivation of the steel. In view of this, the test methods of interest are not just the ones which provide indication for the penetration of the CO_2 gas, but rather those that reflect the penetration and the carbonation reactions that follow. Evaluation of the resistance to penetration of the gas itself might be obtained by using test results of penetration of gases which are nonreactive, such as air and oxygen, as outlined in Section 7.4.

The test methods used to monitor the extent of penetration combined with the carbonation reactions are based on the evaluation of the chemical changes in the concrete as a function of depth from the concrete surface. These could either be characterization of the profile of hydration products as a function of depth (e.g., following the conversion of calcium hydroxide to calcium carbonate) or the change in the nature of the pore solution (e.g., change in the pH as it is reduced from values of about 13.5 to about 10). The two are related, as can be seen in Figure 7.25.

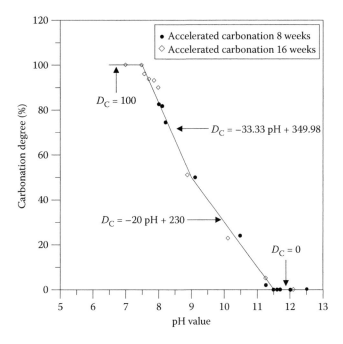

Figure 7.25 Relations between pH and degree of carbonation (% conversion of calcium hydroxide to calcium carbonate). (Adapted from Chang, C.-F. and Chen, J.-W., 2006, *Cement and Concrete Research*, 36, 1760–1767.)

Carbonated concrete
layer

Dashed line added
for contrast

Uncarbonated
concrete

Figure 7.26 Treatment of sawn concrete specimen with phenolphthalein indicator mark-
ing the border line between pore zones with pH below 9 (colorless zones
close to the surface) and above 9 (purple zone at the core). Width scale
100 mm.

It is easiest to follow the carbonation front by using indicator solutions
which change the color as a function of pH. Most common is the
phenolphthalein solution, which marks the boundary for a pH of 9: zones
where the pH is above 9 are colorless and those with pH below 9 are purple
(Figure 7.26).

The zone marked by the phenolphthalein indicator reflects the zone in
which 50% of the conversion of calcium hydroxide due to carbonation has
taken place (X_p in Figure 7.27). Full conversion is at a deeper location,
X_c. From a practical point of view, the X_p depth is of most interest since it
is at this location that the pH of the pore solution has declined to values
that can no longer assure passivation. Thus, the depth X_p marked by the
phenolphthalein indicator is usually noted as the "depth of carbonation."

Carbonation tests are based on the exposure of concrete specimens in the
laboratory to a carbonating environment, in which the humidity, temperature, and CO_2 concentration are controlled to provide accelerated conditions. Periodically, samples are taken out of the carbonating chamber and
the depth of carbonation is determined by the phenolphthalein indicator.
In such tests, all of the faces of the specimen are usually exposed to
carbonation, and the representative depth is taken as the average of all
of the faces except the troweled one, which may have different properties. Guidelines for measuring the depth and averaging are provided in the
RILEM recommendation (RILEM CPC-18, 1988).

The results of the test are presented in terms of depth of carbonation
against time, and usually linear relationships are obtained when using
square root of time.

The carbonation–time curves can thus be described in terms of a linear
relation between the depth of carbonation and the square root of time. The
carbonation rate coefficient can be defined in terms of the following equation:

$$d = K * t^{0.5} \tag{7.7}$$

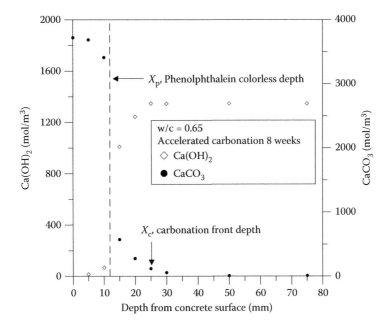

Figure 7.27 Conversion of calcium hydroxide to calcium carbonate as a function of depth of concrete and the location of the carbonation level X_p denoted by phenolphthalein indicator. (Adapted from Chang, C.-F. and Chen, J.-W., 2006, The experimental investigation of concrete carbonation depth, *Cement and Concrete Research*, **36**, 1760–1767.)

where d is the depth of carbonation (mm), t is the exposure time (years), and K is the carbonation rate constant (mm/year$^{0.5}$).

The value of K depends on the carbonation conditions and the quality of the concrete. It can be determined empirically or based on some models in which the carbonating conditions and concrete quality are considered separately. Correlations have been reported between the K values obtained in natural conditions and in accelerated ones.

Accelerated tests are carried out in specially designed chambers and a range of conditions have been reported in literature with temperatures from 30°C to 60°C; relative humidity of 30%–60%RH; and CO_2 concentration over a wide range from 0.3% to 5%.

REFERENCES

ACI Committee 201, 2008, Guide to durable concrete, ACI 201.2R-08, American Concrete Institute.

ACI Committee 365, 2000, Service life prediction—State of the art report, Manual of Concrete Practice, American Concrete Institute.

Alexander, M. G., Mackechnie, J. R., and Ballim, Y., 2006, Use of durability indexes to achieve durable cover concrete in reinforced concrete structures, in *Materials Science of Concrete*, J. P. Skalny and S. Mindess, (eds.), Wiley, Hoboken, NJ, Vol. VI, pp. 483–511.

Alexander, M. G., Mackenzie, J. R., and Ballim, Y., 1999, Guide to the use of durability indexes for achieving durability in concrete structures, Research Monograph No 2, Department of Civil Engineering, University of Cape Town, 35pp.

Ballim, Y., 1991, A low cost falling head permeameter for measuring concrete gas permeability, *Concrete Beton*, 61, 13–18.

Basheer, M., Goncalez, A. F., and Torrent, R., 2007, Non-destructive methods to measure water transport, in *Non-Destructive Evaluation of the Penetrability and Thickness of the Concrete Cover—State-of-the-Art Report of RILEM Technical Committee 189-NEC*, R. Torrent and L. F. Luco (eds.), RILEM Publications SARL, Bagneux, France.

Bentur, A., Berke, N. S., and Li, L., 2006, Integration of technologies for optimizing durability performance of reinforced concretes, in *Concrete Durability and Service Life Planning (ConcreteLife 06)*, K. Kovler (ed.), *Proceedings RILEM International Conference*, Israel, RILEM Publications PRO 46, Paris, pp. 247–258.

Bentur, A., Diamond, S., and Berke, N., 1997, *Steel Corrosion in Concrete*, E&FN SPON/Chapman and Hall, London, UK.

Berke, N. S. and Hicks, M. C., 1992, The life cycle of reinforced concrete decks and marine piles using laboratory diffusion and corrosion data, in *Corrsion Forms and Control for Infrastructure*, V. Chaber (ed.), ASTM STP 1137, American Society of Testing and Materials, West Conshohocken, PA, pp. 207–231.

Bickley, J., Hooton, R. D., and Hover, K. C., 2006, Preparation of performance-based specification for cast-in-place concrete, RMC Research Foundation.

Carino, N. J., 1999, Nondestructive techniques to investigate corrosion status in concrete structures, *Journal of Performance of Constructed Facilities*, August, 96–106.

Carino, N. J., 2004, Methods to evaluate corrosion of reinforcement, in *Handbook on Nondestructive Testing of Concrete*, 2nd edition, V. M. Malhotra and N. J. Carino (eds.), CRC Press, Boca Raton, London, New York, Washington, D.C., Chapter 11.

Chang, C.-F. and Chen, J.-W., 2006, The experimental investigation of concrete carbonation depth, *Cement and Concrete Research*, 36, 1760–1767.

Concrete Society Working Party, 1987, Permeability testing of site concrete—A review of methods and experience, Concrete Society Technical Report No. 31, London, UK.

DARTS, 2004a, Durable and Reliable Tunnel Structures: Deterioration Modelling, European Commission, Growths 2000, Contract G1RDCT-2000-00467, Project GrD1-25633.

DARTS, 2004b, Durable and Reliable Tunnel Structures: Data, European Commission, Growths 2000, Contract G1RD-CT-2000-00467, Project GrD1-25633.

DeSouza, S. J., Hooton, R. D., and Bickley, J. A., 1997, Evaluation of laboratory drying procedures relevant to field conditions for concrete sorptivity measurements, *Cement, Concrete, and Aggregates*, 19(2), 92–96.

Dhir, R. K., Hewlett, P. C., and Chan, Y. N., 1987, Near surface characteristics of concrete: Assessment and development of *in situ* test methods, *Magazine of Concrete Research*, 39(141), 183–195.

DuraCrete, 2000, Probabilistic performance based durability design of concrete structures: Statistical quantification of the variables in the limit state functions, Report No. BE 95-1347.

El-Dieb, A. S. and Hooton, R. D., 1995, Water permeability measurement of high performance concrete using a high pressure triaxial cell, *Cement and Concrete Research*, 25(6), 1199–1208.

Figg, J. W., 1973, Methods of measuring the air and water permeability of concrete, *Magazine of Concrete Research*, 25(85), 213–219.

Figg, J. W., 1988, Discussion of the paper R. K. Dhir, P. C. Hewlett, and Y. N. Chan, Near surface characteristics of concrete: Assessment and development of in situ test methods, in *Magazine of Concrete Research* 1987, 39(141) 183–195, *Magazine of Concrete Research*, 40(145), 234–244.

Hamilton, R. H. and Boyd, A. J., 2007, Permeability of concrete—Comparison of conductivity and diffusion methods, Research Report, UF Project No. 00026899, Civil and Coastal Engineering, University of Florida.

Kollek, J. J., 1989, The determination of the permeability of concrete to oxygen by the CEMBUREAU method—A recommendation, *Materials and Structure*, 22, 225–230.

Lobo, C. L., 2007, New perspective on concrete durability, *Concrete*, Spring, 24–30.

Medeiros, M. H. F. and Helene, P., 2009, Surface treatment of reinforced concrete in marine environment: Influence on chloride diffusion coefficient and capillary water absorption, *Construction and Building Materials*, 23, 1476–1484.

Otzsuki, N., Nagataki, S., and Nakashita, K., 1992, Evaluation of $AgNO_3$ solution spray method for measurement of chloride penetration into hardened cementitious matrix materials, *ACI Materials Journal*, 89(6), 587–592.

RILEM Committee TC 116, 1999, Permeability of concrete as a criterion of its durability, *Materials and Structures*, 32, 174–179.

RILEM Recommendation CPC-18, 1988, Measurement of hardened concrete carbonation depth, *Materials and Structures*, 21(6), 453–455.

Romer, M., 2005, Effect of moisture and concrete composition on the Torrent permeability measurement, *Materials and Structures*, 38, 541–547.

Tang, L. and Nilsson, L.-O., 1992, Rapid determination of the chloride diffusivity in concrete by applying an electrical field, *ACI Materials Journal*, 89(1), 49–53.

Torrent, R., Basheer, M., and Goncalez, A. F., 2007, Non-destructive methods to measure gas-permeability, in *Non-Destructive Evaluation of the Penetrability and Thickness of the Concrete Cover—State-of-the-Art Report of RILEM Technical Committee 189-NEC*, R. Torrent and L. F. Luco (eds.), RILEM Publications SARL, Bagneux, France.

Torrent, R. and Luco, L.F. (eds.), 2007, *Non-Destructive Evaluation of the Penetrability and Thickness of the Concrete Cover—State-of-the-Art Report of RILEM Technical Committee 189-NEC*, RILEM Publications SARL, Bagneux, France.

Torrent, R. and Frenzer, G., 1995, Methoden zur messing und Beurteilung der Kennwerte des Uberdeckungsbetons auf der Baustelle—Teil 2 Office Federal de Routes, Suisse, Rapport No. 516, Zurich, Oktober, 106pp.

Torrent, R. J., 1992, A two-chamber vacuum cell for measuring the coefficient of permeability to air of the concrete cover on site, *Materials and Structures*, 25(6), 358–365.

Wasserman, R. and Bentur, A., 2013, Efficiency of curing technologies: Strength and durability, *Materials and Structures*, 46, 1833–1842.

Wong, H. W. and Saraswathy, V., 2007, Corrosion monitoring of reinforced concrete structures—A review, *International Journal of Electrochemical Science*, 2, 1–28.

Chapter 8

Durability testing

Degradation mechanisms

8.1 INTRODUCTION

This chapter deals with durability tests intended to characterize the distress caused by the interaction of penetrating deleterious agents with concrete components, leading to deterioration of the concrete itself or of the reinforcing steel embedded in it. Such tests are intended to assess the nature and the level of damage that may be caused, as well as the choice of materials to be used to minimize such effects. These tests also serve the research and development of new materials and systems of enhanced durability performance. In some cases, the tests are used as a basis for specifications by providing limiting values on materials to be applied in structures exposed to various classes of environmental exposures.

Frequently, the tests involve accelerating techniques such as cycling of weathering effects. Investigations have been carried out to correlate the results of the accelerated tests with natural exposure.

Overviews of these tests are presented in many references, and those that have been particularly used here are ACI (2000, 2008), Andrade et al. (2007), Bentur et al. (1997), Bickley et al. (2006), Duracrete (2000), Darts (2004a,b), Lobo (2007), Carino (1999, 2004), Sagues (1993), Wong (2007), Lindgard et al. (2012), Torrent and Luco (2007), and RILEM TC116 (1999a,b).

8.2 TESTS IDENTIFYING AND QUANTIFYING DEGRADATION PROCESSES OF CONCRETE

Numerous tests have been developed over the years to identify degradation processes and quantify their influence. The most common processes addressed in such tests include sulfate attack, alkali–aggregate reactions and frost attack. The tests are mainly intended to identify the risks involved in the various concrete ingredients (e.g., susceptibility of aggregates to alkali attack and cements to sulfate attack) and the effectiveness of various means to provide resistance to the attack (e.g., pozzolans to mitigate alkali aggregate reaction and sulfate attacks; air to provide frost resistance). The tests can provide the basis for determination of the quantitative parameters needed for guidelines for specification and use of various materials and technologies in terms of the risks involved. They are particularly useful for comparative purposes.

The common feature of all of these tests is the quantification of the distress induced by the attack by measuring the overall damage that occurs. These show up in behaviors such as length change and reduction in modulus of elasticity of the tested specimens. These characteristics are determined as a function of time. The specifications set for passing the test require that the extent of damage should not exceed a critical value within a specified time of testing or number of cycles applied. The tests are designed to have in them accelerating factors such as intensive cycling of freeze–thaw and exposure to very concentrated sulfate solutions. Even with these means, they often take quite a long time, several months in duration.

8.2.1 Alkali–aggregate reaction tests

Alkali–aggregate tests are intended not only to quantify the risk involved in the aggregates used, but also the influence of the nature of the cement, particularly in view of its alkali content. The outcome of the tests is used for specifications of aggregates, such as in ASTM C33 "Standard Specification for Concrete Aggregates." This standard refers to the various test methods but does not provide guidance to a preferred test.

The quality of aggregates in combination with the cement can be evaluated by ASTM C227, ASTM C287, ASTM C289, ASTM C1260, ASTM C1105, and ASTM C1293. They include tests based on graded and ground aggregates evaluated in mortars (ASTM C227 and ASTM C1260), full size aggregates in concrete (ASTM C1105 and ASTM C1293), chemical tests to evaluate the reactivity of the aggregates in solution (ASTM C289), and special tests to assess the effectiveness of supplementary cementitious materials to reduce the alkali attack (ASTM C441 and ASTM C1567).

The evaluation of the deleterious reaction with cement is based on the measurement of length changes. In addition to length change measurements, it is recommended to provide qualitative assessment of the damage in terms of the extent of warping, cracking, surface mottling, and deposits or exudations, including their nature and thickness. The deleterious reaction in these tests is often accelerated by means of exposure to high temperatures. The tests can be classified into mortar tests, concrete tests, chemical tests of the stability of the aggregates, and tests for assessing the effectiveness of supplementary cementitious materials.

   a. *Mortar tests*:
      • Test at 38°C in water: ASTM C227 "Potential Alkali Reactivity of Cement–Aggregate Combinations (Mortar-Bar Method)" using cement of the highest alkali content, which may be used in practice for the tested aggregate.

- Test at 80°C in NaOH solution: ASTM C1260 "Potential Alkali Reactivity of Aggregates (Mortar-Bar Method)." This test uses a highly alkaline solution to screen for deleterious or non-deleterious aggregates. The criterion set in the ASTM C1260 test states that expansion of less than 0.10% at 16 days after casting is indicative of innocuous behavior, while expansion of more than 0.20% at 16 days after casting is indicative of potentially deleterious expansion.

b. *Concrete tests*:
  - ASTM C1105 "Length Change of Concrete Due to Alkali–Carbonate Rock Reaction" requires conditioning of the concrete with the tested aggregate in moist conditions, but not immersed in water, using cement having the highest alkali content representative of the general use intended, or available to the laboratory making the tests. It is commented in the standard that a cement–aggregate combination might reasonably be classified as potentially deleteriously reactive if the average expansion is equal to or greater than 0.015% at 3 months; 0.025% at 6 months; or 0.030% at 1 year.
  - ASTM C1293 "Determination of Length Change of Concrete Due to Alkali–Silica Reaction" is based on adding NaOH to the mix water so that the total equivalent alkali content in the mix is 1.25% by weight of cement; the concretes are stored in sealed conditions at 38°C. These are accelerated conditions in which length changes are monitored. The standard does not provide recommendations for acceptable expansion values and refers back to the ASTM C33 standard, which sets the requirements for the quality of aggregates.

c. *Chemical test of aggregates*:
  - ASTM C289 "Potential Alkali–Silica Reactivity of Aggregates (Chemical Method)" is chemical in nature and measures the dissolved silica and reduction in alkalinity of ground aggregate in NaOH solution at 80°C. Some guidelines for classifying aggregate as susceptible are provided in Figure 8.1.

d. *Effectiveness of supplementary cementitious materials*:
  - The effectiveness of supplementary cementitious materials in mitigating alkali attack on aggregates can be evaluated by means of the ASTM C441 and ASTM C15672 tests.
  - ASTM C441 "Effectiveness of Mineral Admixtures or Ground Blast-Furnace Slag in Preventing Excessive Expansion of Concrete Due to the Alkali–Silica Reaction" is based on testing the expansion of mortars prepared with ground Pyrex aggregates (highly reactive amorphous silica) in a 38°C wet environment. The binder is made from Portland cement having an alkali content in the range of 0.95%–1.05% by weight of cement and a mix in which

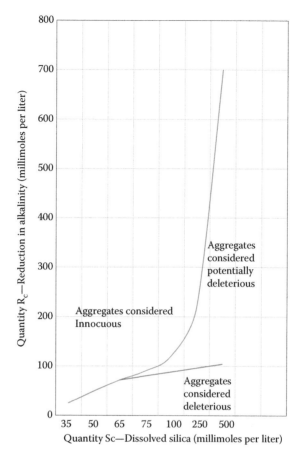

Figure 8.1 Illustration of division between innocuous and deleterious aggregates on the basis of reduction in alkalinity test. (Adapted from ASTM C289.)

2.5% of the volume of the cement has been replaced by the mineral admixture. The effectiveness of the mineral admixture is quantified in terms of the relative reduction in the expansion.

- ASTM C1567 "Standard Test Method for Determining the Potential Alkali–Silica Reactivity of Combinations of Cementitious Materials and Aggregate (Accelerated Mortar-Bar Method)" is similar in terms of the specimens prepared to ASTM C1260 (immersion of mortar in NaOH solution at 80°C). It evaluates the reduction in expansion using mortars with the tested aggregate due to the incorporation of pozzolans and ground-granulated blast-furnace slag as partial replacement of the cement. The test is based on the accelerated test method developed at the National Building Research Institute (NBRI) in South Africa. It allows the detection within 16 days of the potential for deleterious alkali–silica reaction of combinations of cementitious materials and aggregate in mortar bars and its mitigation using various proportions of hydraulic cement, pozzolans, and ground-granulated blast-furnace slag.

Lindgard et al. (2012) reviewed the test methods and made some comments on the issues that need to be considered in the mix design used for the evaluation of the aggregates for susceptibility to alkali–aggregate attack:

- Water/binder ratio has a dual effect: decreasing the ratio may result in higher OH⁻ ion concentration leading to enhanced attack while at the same time restraining expansion due to the higher strength matrix.
- Alkali content is sometime boosted to accelerate the test and this may mask its role on the actual job site.
- Alkali leaching should be minimized by having appropriate storage conditions and specimen sizes: reduced leaching occurs with larger specimen sizes, lower temperature, and less moisture condensation on the specimens.
- The laboratory test should be designed to expose the specimens to worst-case humidity conditions while at the same time making sure that they will not lead to leaching of alkalis.
- When using high-strength, low w/c ratio concretes, internal desiccation can take place and this should be taken into account to avoid a too low internal RH in the concrete.
- Acceleration of the test by an increase in the temperature over 38°C should be avoided as it may generate effects that do not reflect the room temperature processes.
- The curing before the test is of particular importance when pozzolanic materials are used, to allow for sufficient reaction to take place prior to accelerated exposure.

8.2.2 Sulfate resistance tests

The sulfate reaction results in expansion and therefore testing specimens for expansion is the basis for test methods to evaluate sulfate resistance, such as ASTM C452 "Test Method for Potential Expansion of Portland-Cement Mortars Exposed to Sulfate" and ASTM C1012 "Test Method for Length Change of Hydraulic-Cement Mortars Exposed to a Sulfate Solution." The sulfate attack is usually an external one, induced by sulfate ions penetrating from the surface inward. Penetration into the concrete can be quite slow in concrete specimens, which are several tens of millimeters in cross section. Thus, sulfate resistance tests of concrete specimens can take quite long, in contrast to tests of alkali–aggregate reaction, which are much shorter in time since the degrading agents are within the concrete itself. As a result, many of the sulfate resistance tests are based on mortar specimens, which can be made smaller in size.

In order to accelerate the sulfate test, ASTM C452 is based on the production of specimens in which 7% sulfate is introduced into the mix by incorporating the appropriate quantity of gypsum. This enables immediate and accelerated interactions to determine the performance of the cement used within 14 days of monitoring of expansion.

This test is not adequate for blended cements where the mitigating mechanism involves the progress of the pozzolanic reaction to produce hydration products and microstructure, which provide sulfate resistance. Since sulfate is available immediately, there is no time for these internal changes to occur in time to affect the sulfate attack. For that purpose, ASTM C1012 is more appropriate since it measures the resistance by immersion in sodium sulfate solution after 20 MPa strength has been attained. This implies that sulfate ingress is initiated only after pozzolanic reactions have taken place, which better represents the reality. Results of evaluations using this test, demonstrating the positive influence of mineral additives, are presented in Figure 8.2. Requirements specified by ACI with regard to the results obtained from the ASTM C1012 test are presented in Table 8.1.

It should be noted that the immersion test does not represent the harsh conditions that may occur in practice where wetting and drying of sulfate-containing water may result in the buildup of sulfate concentrations, which

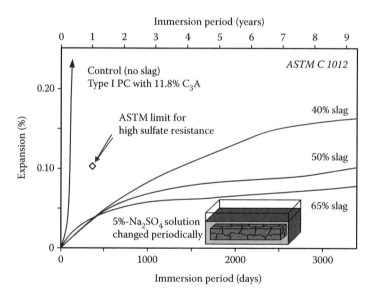

Figure 8.2 Effect of mineral admixtures on sulfate resistance of mortars tested according to the ASTM C1012 test. (From Hooton, R. D. and Emery, J. J., 1990, Sulfate resistance of a Canadian slag cement, *ACI Materials Journal*, 87(6), 547–555. With permission.)

Table 8.1 Limit of ACI 201 code on the resistance to sulfate in various environmental condition specified in terms of the ASTM C1012 expansion test

Exposure severity	Expansion in the ASTM C1012 test
Class 0—Negligible	
Class 1—Moderate	0.10% at 6 months
Class 2—Severe	0.05% at 6 months or 0.10% at 12 months
Class 3—Very severe	0.10% at 18 months

are far larger than those simulated in the immersion test. Also, the immersion in solution may lead to changes in the pH, which may not reflect real-life conditions. There is a debate going on as to what effect this has on such tests and their interpretation.

The acceleration provided in the standard tests reviewed above is not always sufficient for laboratory testing, and so alternatives have been used both in research and in evaluations of sulfate-resistant systems. These are usually based on wetting and drying of the specimens in sulfate solutions such as sodium sulfate, which simulates the action observed in exposure to marine environments and contaminated soils. The ingress of the sulfate into the concrete is driven in these tests by capillarity, which is a much more intense penetration process than the diffusion in continuously immersed solutions. Wetting and drying can, however, result also in the crystallization of the salt in the pores, creating a pressure associated with the increased volume occupied by the crystals, which can further accelerate the deterioration process. This kind of effect may simulate the processes taking place in the tidal zone in marine structures, which is the part of the structure where the exposure conditions are the harshest.

8.2.3 Freeze–thaw resistance

The resistance to freeze–thaw cycles is dependent on the strength achieved by the concrete and the quality of the air-void system induced into it. The aim is to have the average bubble spacing not exceeding 230 µm, with no single value over 260 µm. This can be tested by microscopic means following the ASTM C457 test method for "Microscopical Determination of Parameters of the Air-Void System in Hardened Concrete."

The physical testing of the resistance to freeze–thaw cycles requires careful conditioning of the moisture content in the concrete. The ASTM C666 test "Method for Resistance of Concrete to Rapid Freezing and Thawing" specifies two types of conditioning: (i) Procedure A in which both freezing and thawing are carried out with the specimen submerged in water and (ii) Procedure B where the freezing is in air and thawing is in water. These tests provide an indication as to the effects of concrete composition but cannot serve to estimate service life. The damage induced in the cycles is periodically monitored to determine internal damage by evaluation of the changes in dynamic modulus of elasticity and length.

The amount of surface scaling can be measured by determining the weight loss. In the European CIF (Capillary suction-Internal damage-Freeze/thaw resistance) test (Setzer et al., 2001), concrete slabs are dried in air at 65%RH for 21 days and then saturated by capillary absorption for 7 days, followed by exposure to the freezing and thawing cycles. Scaled material is removed in an ultrasonic bath and weighed, while internal damage is assessed by determining the ultrasonic pulse velocity.

The resistance to freezing and thawing in the presence of deicing salts can be assessed by the ASTM C672 test "Scaling Resistance of Concrete Surfaces Exposed to Deicing Chemicals." (This test was withdrawn in 2012.) The concrete surface is brushed as a final finishing operation. After curing, the specimen is covered with a calcium chloride solution and exposed to freezing and thawing cycles. The monitoring of damage is visual and thus qualitative in nature.

Testing of the actual freeze–thaw resistance of the aggregates can be done by subjecting them to wetting and drying cycles in salt solutions with the resulting crystallization of the salt in the pores of the aggregates leading to expansion and cracking, simulating the action of freezing water. The ASTM C88 test "Soundness of Aggregates by Use of Sodium Sulfate or Magnesium Sulfate" is based on this principle. It uses sodium sulfate or magnesium sulfate solutions, where the wetting cycle consists of immersion in the solution for 16–18 hours at 21°C and the drying is carried out at 110°C until weight equilibrium is achieved. The required number of cycles and limits on weight loss are not set in the standard; they are to be specified by the relevant agency requiring the test. Usually, the weight loss limits are set at 12% and 18% for sodium sulfate and magnesium sulfate, respectively. The standard also recommends qualitative characterization after the cycling, looking into splitting, crumbling, cracking, and flaking.

8.3 CORROSION OF STEEL IN CONCRETE

8.3.1 Introduction

The electrochemical processes involved in steel corrosion were described in detail in Chapter 4. The degradation processes associated with corrosion of steel in concrete require monitoring of the actual state of the steel, whether by laboratory tests or by site monitoring. Observation of the steel bars themselves by measurement of the reduction in their diameter or weight loss is a destructive test, which is obviously difficult to perform and is thus very limited in practice. Evaluation of the state of the steel therefore requires nondestructive tests. These are based on electrical monitoring, which is made feasible by the fact that corrosion of steel in concrete is an electrochemical process involving changes which can be monitored by electrical signals. The signals can be of various types, providing indirect indications of whether there is a potential for the corrosion process to take place (i.e., whether depassivation has occurred) or more direct measurements of the corrosion rate itself in systems where corrosion is ongoing. The main tests will be reviewed here and discussed within the context of their significance and limitations. The review is based on the following references: Sagues (1993), Bentur et al. (1997), Carino (1999, 2004), and Wong and Saraswathy (2007).

8.3.2 Electrochemical corrosion and nondestructive testing techniques

The driving forces for creating the electrochemical process of corrosion can be presented in electrical terms, which form the basis for various nondestructive testing techniques. When the steel is in the active corrosion state, that is, depassivation has occurred, it undergoes a change in its potential, and potential differences are set up between anodic and cathodic sites on the surface of the steel. Anodic sites are where depassivation has occurred while at the cathodic sites passivation is still effective. At this stage, a galvanic cell is set up and current flows between the cathodic and anodic sites (see Chapter 4) resulting in changes in the potential of each of the sites until equilibrium is reached, as shown schematically in Figure 8.3. The curves of the change in potential with increase in the current are the polarization curves characterizing the behavior at

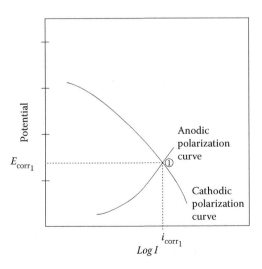

Figure 8.3 Polarization curves in corrosion process simulated by short-circuited cell, showing the corrosion potential E_{corr} and the corrosion current i_{corr}.

the anode and cathode. The equilibrium reached at the intersection of the curves represents the steady-state corrosion process, which is characterized by the corrosion current, i_{corr}, and the corrosion potential, E_{corr}. The corrosion current is a direct measure of the rate of corrosion, and from it one may calculate the rate of loss of steel in units of weight, diameter, and cross section.

In a steel system which undergoes passivation, the anodic polarization curve assumes a different form: a slight increase in current and potential initially, representing an oxidation reaction, and thereafter a sharp decline in the current with increase in potential, representing the formation of the passivation film and the isolation of the steel from its surrounding, causing the sharp decline in the current to negligible values (Figure 8.4a).

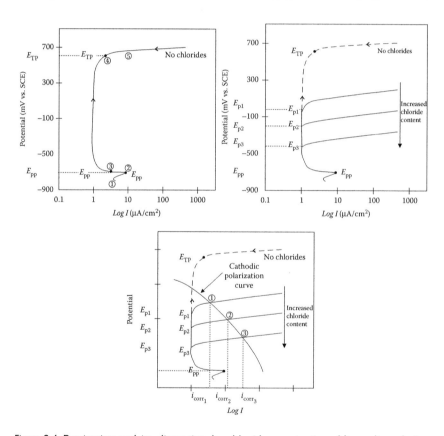

Figure 8.4 Passivation and its disruption by chloride penetration: (a) anodic polarization curve of passivated system, (b) anodic polarization curves demonstrating the breakdown of passivation with increasing content of chlorides, and (c) corrosion current and potential affected by increase in chloride content as determined from the intersection of the anodic and cathodic polarization curves. (After Bentur, A., Diamond, S., and Berke, N., 1997, *Steel Corrosion in Concrete*, E&FN SPON/Chapman and Hall, London, UK.)

Depassivation by the penetration of chloride shows up as a sharp rise in current above a certain critical potential value, which is related to the chloride content. A higher chloride content results in a lower depassivation potential (Figure 8.4b) and higher corrosion current, as demonstrated in Figure 8.4c for the intersection of the anodic and cathodic polarization curves.

The electrical parameters outlined above (corrosion potential, corrosion current, and electrical resistance) can be characterized, and their interpretation by expert analysis can serve as the basis for a range of nondestructive tests for assessing the corrosion behavior in the laboratory as well as on site. Tests to determine the electrical parameters can assist in characterizing the state of corrosion, whether the steel is in the active or passivated state, as well as determine the rate of corrosion when in the active state. Such tests will be reviewed in this section. Penetration tests of chloride ingress and carbonation, such as those outlined in Chapter 7, can serve to quantify the quality of the concrete cover with respect to the time until depassivation. This has often been used to estimate the service life of the reinforced system.

8.3.3 Half-cell potential method

Measurement of the potential of the steel embedded in the concrete can provide an indication of whether it is in the active state. The half-cell potential measurement can be used for this purpose, as illustrated in Figure 8.5.

The test method, ASTM C876 "Half-Cell Potentials of Uncoated Reinforcing Steel in Concrete," uses a copper–copper sulfate half-cell (the reference electrode which is made up of a copper rod immersed in a copper sulfate solution) and is in electrical contact with the surface of the concrete (achieved by means of a porous plug and a sponge which is moistened with a wetting solution). The half-cell electrode is affected by the potential in the concrete cover which is induced by the excess electrons flowing from the corroding bar. The potential difference between the copper–copper half-cell and the reinforcing bar measured by a high-impedance voltmeter (connected on one side to the bare bar and the other side to the half-cell electrode) is expected to become more negative as the bar corrodes since more excess electrons become available.

A survey of the potentials measured by this method can be expressed in terms of an equipotential map, as shown in Figure 8.6.

Guidelines for interpretation of the test results usually suggest that when the potential is higher (i.e., more positive) than −200 mV it is unlikely that the steel bar is active and corrosion is unlikely to occur. When it is below −350 mV (i.e., more negative), it is likely that the steel bar is active, implying that depassivation has taken place and corrosion may be occurring.

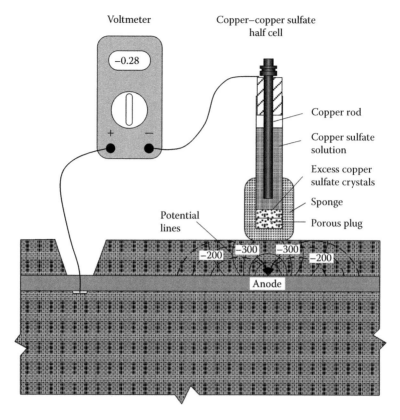

Figure 8.5 Test setup for half-cell potential measurements to determine the surface potential associated with the corrosion current, ASTM C876. (Adapted from Carino, N. J., 2004, Methods to evaluate corrosion of reinforcement, in *Handbook on Nondestructive Testing of Concrete*, 2nd edition, V. M. Malhotra and N. J. Carino (eds.), CRC Press, Chapter 11, Copyright 2004, reproduced by permission of Taylor & Francis Group, LLC, a division of Informa plc.)

The results of this test should be viewed as qualitative in nature and should be interpreted with care, due to some limitations as outlined below:

- Even if the steel bar is depassivated, the corrosion rate might be low since it depends on factors such as the availability of oxygen and the electrical resistivity of the concrete. Relations reported between half-cell potential and corrosion rates are therefore not clear-cut. Corrosion rates above 1 $\mu A/cm^2$ can be considered to represent high corrosion while values below 0.1 $\mu A/cm^2$ are indicative of negligible corrosion rates.
- It should also be noted that the actual potential measured on the surface of the concrete is not really the same as that of the concrete in contact with the steel (Figure 8.7). The distribution of the potential on

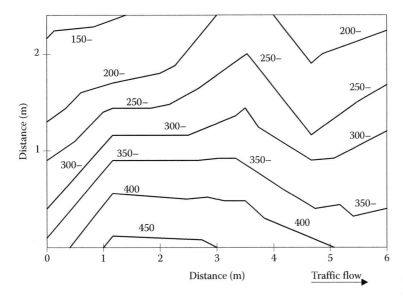

Figure 8.6 Example of equipotential contour map for bridge deck. (After Bentur, A., Diamond, S., and Berke, N., 1997, *Steel Corrosion in Concrete*, E&FN SPON/ Chapman and Hall, London, UK.)

the concrete surface is broader than that in contact with the steel, and this broadening depends on the thickness of the cover and the electrical resistivity of the concrete. Such broadening may detract from the quality of the resolution of the test.

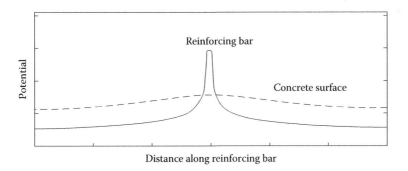

Figure 8.7 Schematic description of potential distribution on the concrete surface and reinforcing bar. (Adapted from Sagues, A. A., 1993, Corrosion measurement techniques for steel in concrete, Paper No. 353 in Corrosion 93, *The NACE Annual Conference and Corrosion Show*; Kranc, S. C. and Sagues, A. A., 1992, Computation of corrosion macrocell current distribution and electrochemical impedance of reinforcing steel in concrete, in *Computer Modeling in Corrosion*, ASTM STP 1154, R.S. Munn (ed.), American Society of Testing and Materials, Philadelphia, PA, p. 95.)

- The relationships with potential become very complicated and not well established when the steel bar is coated with epoxy or galvanized and when the concrete around the bar is carbonated. Under these conditions, there is a special need for evaluation of the results by a specialist. The ASTM standard calls for such precautions.

8.3.4 Concrete resistivity

The electrical resistivity of the concrete cover is an important parameter controlling the rate of corrosion reactions once depassivation has taken place. High resistivity will slow down these reactions and reduce them to values which might be considered as insignificant. Thus, resistivity measurements can provide a complementary input to potential measurements, which are indicative of the likelihood for corrosion to occur.

The resistivity is dependent on the pore structure and moisture state of the concrete. Higher resistivity values are obtained in denser and drier concretes. Thus, measurement on concrete that is wetted can serve as an indication for its quality, while for a known grade of concrete the resistivity can provide an input with regards to its moisture content.

Resistivity (ρ) is defined in terms of the dimensions (length, L; and cross-sectional area, A) and resistance (R) of a conductor:

$$R = \rho \left(\frac{L}{A} \right) \tag{8.1}$$

The common test used to determine the resistivity of concrete cover is the Wenner test (four-point test), shown schematically in Figure 8.8, which is based on four equally spaced electrodes.

Alternating electric current is impressed on the outer electrodes and the resulting current is measured between the inner electrodes. The lines of equipotential and equicurrent are shown in Figure 8.8 and the calculation of the resistivity is based on the Wenner equation:

$$\rho = 2 \cdot \pi \cdot R \cdot s \tag{8.2}$$

The resistivity calculated is an "apparent" one because the Wenner equation assumes that the material is semi-infinite and homogeneous and that the electrodes are point shaped (infinitely small) and placed on the surface. In order to obtain an estimate in which the deviations from the "real" resistivity are as small as possible, the following needs to be considered:

- The spacing between the electrodes should be significantly larger than the maximum size of the aggregate.

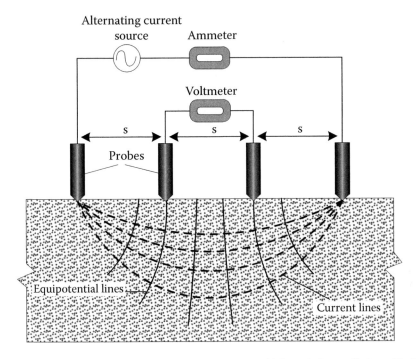

Figure 8.8 Four-probe Wenner test for resistivity. (Adapted from Carino, N. J., 2004, Methods to evaluate corrosion of reinforcement, in *Handbook on Nondestructive Testing of Concrete*, 2nd edition, V. M. Malhotra and N. J. Carino (eds.), CRC Press, Chapter 11, Copyright 2004, reproduced by permission of Taylor & Francis Group, LLC, a division of Informa plc.)

- The spacing should not be too large relative to the concrete cross section in order to eliminate boundary effects.
- Measurements made too close to the steel bars can lead to erroneous results.

In view of these constraints, recommendations for positioning of the probes of the four-point Wenner test have been suggested as a function of the maximum aggregate size, d_{max}, and the slab thickness, h (Gowers and Millard, 1999; Broomfield and Millard, 2002; Angst and Elsener, 2014). The probe spacing should be greater than $1.5\ d_{max}$ and smaller than $0.5\ h$ to achieve a measurement with less than 10% error, and less than 0.25–0.33 of the slab thickness h, to achieve a measurement with less than 5% error. The distance from the slab edge should be greater than twice the spacing between the probes.

Angst and Elsener (2014) noted that the Wenner four-point measurement on the surface yields values that are smaller than those obtained in the bulk concrete. They attributed this difference to the influence of coarse aggregates, which distort the current field and thus affect the measurement, especially at the sensing electrode. Angst and Elsener thus concluded that the four-point surface measurement cannot reproduce the bulk concrete values. Yet they noted that the difference is 10% or even much less if the guidelines outlined above are followed.

In addition, precautions should be taken that the measurements are not affected by the presence of reinforcing bars and for that purpose they should be made midway between reinforcing bars.

8.3.5 Linear polarization resistance

The most direct and valuable way of quantifying the state of corrosion of the steel in concrete is the evaluation of the corrosion current in the reinforcement when depassivation has taken place. In the equilibrium corroding state, the corrosion current indicates the rate of corrosion. However, this is an internal current within the system and cannot be directly measured. Yet, it has been shown that by a small external intervention (change in potential or impressed current) the equilibrium point is shifted slightly and a linear relation can be established between the changes in the potential, ΔE, and the current, Δi, which is related to the corrosion current i_{corr}. This range, known also as the Tafel extrapolation range, has been quantified by Stern and Geary (1957) in terms of the polarization potential, R_p:

$$R_p = \frac{\Delta E}{\Delta i} \tag{8.3}$$

They showed that R_p is related to the corrosion current, i_{corr}:

$$i_{corr} = \frac{B}{R_p} \tag{8.4}$$

where B is a constant with a value of 26 mV and R_p is in units of ohms times area, and therefore it is not a true resistance although the term polarization resistance is commonly used. The linear curve obtained in such a test is shown in Figure 8.9.

On the basis of this concept, the linear polarization test method was developed, as shown schematically in Figure 8.10.

The system consists of three electrodes, one being the reference half-cell, the other the reinforcing bar itself (the working electrode), and the third one being the counter electrode which supplies the current. The system is equipped with devices to determine the current and the potential difference. It can operate in two modes:

- The potentiostatic mode in which the current is changed to provide a constant potential of the working electrode

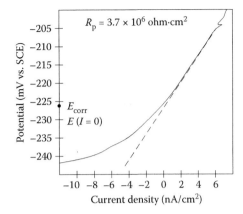

Figure 8.9 Polarization resistance curve. (Adapted from Bentur, A., Diamond, S., and Berke, N., 1997, *Steel Corrosion in Concrete*, E&FN SPON/Chapman and Hall, London, UK; after Berke, N. S. and Hicks, M. C., 1990, Electrochemical methods of determining the corrosivity of steel in concrete, *ASTM STP 1000, 25th Anniversary Symposium Committee GI*, American Society of Testing and Materials, Philadelphia, PA, pp. 425–440.)

- The galvanostatic mode in which the potential is changed to provide a constant current from the counter electrode to the working electrode

Open circuit potential, that is, the half-cell potential, can be measured by setting the counter electrode switch open (Figure 8.10a). It is recommended to apply steps of voltage change of 4 mV for the polarization measurement. The affected area of the working electrode is assumed to be the surface area of the reinforcing bar just below the reference electrode. To assure this assumption, a modification to the test method can be adopted using an auxiliary electrode called a guard electrode (Figure 8.11). The guard electrode is kept at a constant potential equal to that of the counter electrode.

Special guidelines and precautions that need to be taken in performing the test were outlined by Carino (2004):

- The concrete surface has to be smooth (not cracked, scarred, or uneven).
- The concrete surface has to be free of water, impermeable coatings, or overlays.
- The cover depth has to be less than 100 mm.
- The reinforcing steel cannot be epoxy coated or galvanized.
- The steel to be monitored has to be in direct contact with the concrete.
- The reinforcement is not cathodically protected.
- The reinforced concrete is not near areas of stray electric currents or strong magnetic fields.

(a)

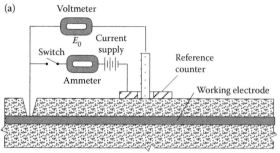

Measure open circuit potential, E_0

(b)

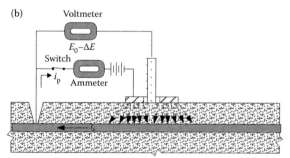

Measure current i_0 to produce small change in voltage ΔE

Figure 8.10 Three-electrode system for linear polarization resistance tests: (a) Measure open circuit potential, E_0 and (b) measured current i_p to produce small change in voltage ΔE. (Adapted from Carino, N. J., 2004, Methods to evaluate corrosion of reinforcement, in *Handbook on Nondestructive Testing of Concrete*, 2nd edition, V. M. Malhotra and N. J. Carino (eds.), CRC Press, Chapter 11, Copyright 2004, reproduced by permission of Taylor & Francis Group, LLC, a division of Informa plc.)

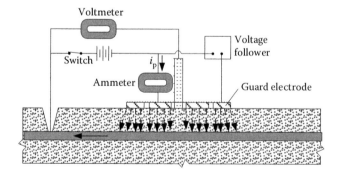

Figure 8.11 Linear polarization device using a guard electrode. (Adapted from Carino, N. J., 2004, Methods to evaluate corrosion of reinforcement, in *Handbook on Nondestructive Testing of Concrete*, 2nd edition, V. M. Malhotra and N. J. Carino (eds.), CRC Press, Chapter 11, Copyright 2004, reproduced by permission of Taylor & Francis Group, LLC, a division of Informa plc.)

- The ambient temperature is between 5°C and 40°C.
- The concrete surface at the test location must be free of visible moisture.
- Test locations must not be closer than 300 mm to discontinuities, such as edges and joints.

Guidelines for assessing the corrosion risks based on the electrical current measurements were proposed by Carino (2004), with classification of corrosion risk into four categories: negligible, low, moderate, and high for corrosion current intensities (μA/cm^2 units) in the ranges of <0.1, 0.1–0.5, 0.5–1.0, and >1 μA/cm^2, respectively.

The rates of corrosion, which are measured in units of current density, can be re-calculated to provide values of loss of weight using Faraday's law (ASTM G102), to express them in terms of steel bar diameter reduction. This calculation provides values that are of direct engineering significance. A 1 μA/cm^2 current corresponds to 0.012 mm/year reduction in steel bar diameter. It should be noted, however, that in a calculation of this kind it is assumed that the corrosion is uniform across the bar. This is the case in carbonation-induced corrosion but not in chloride corrosion where the degradation process may be localized and the values for reduction in the steel cross section might be considerably higher than the calculated average, by a factor as large as one order of magnitude.

It should also be noted that the corrosion rate measurement reflects the rate at the time of measurement and this is very sensitive to the conditions at the time of test, such as the moisture content and temperature of the concrete. These conditions can significantly affect the rates of corrosion of active steel bar. Therefore, one should not rush to draw conclusions based on a single test but rather make sure that periodic measurement are taken over the period of a year, to obtain values which reflect the changes in the environmental conditions.

8.3.6 Electrochemical impedance spectroscopy

The electrochemical impedance spectroscopy (EIS) technique is based on applying a sine or cosine wave of AC current with magnitude i_0 and frequency f as input. The potential (voltage) response is recorded and characterized by the magnitude of V_f and a phase angle of ϕ_f, with respect to the current, as shown schematically in Figure 8.12. The measurements of the amplitude and phase difference of the excitation current and the potential response are carried out at various frequencies, usually in the range of 1 mHz to 1 kHz. At each frequency, the electrochemical impedance is defined as a complex vector equal to the ratio of the potential response to the excitation current.

The response at the different frequencies can be described in terms of processes modeled by an electrical analogue consisting of resistance, capacitance, and/or inductance, as shown in Figure 8.13 for the simplest description.

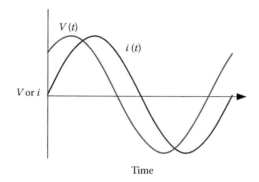

Figure 8.12 Excitation current and resulting potential in the EIS method.

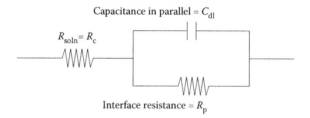

Figure 8.13 Equivalent circuit used to model electrochemical processes at the interfaces.

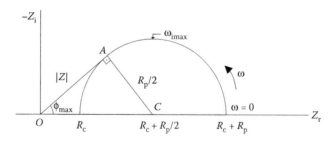

Figure 8.14 Nyquist plot.

In this model, the resistance of the concrete is represented by R_c, the interfacial resistance (also referred to as polarization resistance) is represented by R_p, and the double layer at the steel–concrete interface is represented by the capacitance (C_{dl}) in parallel with R_p.

The impedance using this model can be described in terms of real (resistance) and imaginary (capacitive or induction) components Z and Z'', respectively. They can be obtained by plotting the imaginary impedance against the real one; the result is a semicircle with a diameter of R_p and which is offset by a value of R_c, as shown schematically in Figure 8.14.

The characteristic values obtained in this plot can be used to model the system in terms of the ohmic resistance of the steel cover (between the reference half-cell and the reinforcing bar) and the double-layer capacitance, as shown in Figure 8.13.

R_c is the concrete cover ohmic resistance.

The double-layer capacitance value, C_{dl}, can be calculated as

$$C_{dl} = \frac{1}{2} \cdot \pi \cdot R_p \cdot f \qquad (8.5)$$

where f is the frequency at the highest point on the semicircle.

Thus, the EIS method is more informative because it can provide assessment of the processes at the steel/concrete interfaces and not just the polarization resistance R_p. However, the test is very time consuming and therefore it has been used mainly for research.

REFERENCES

ACI, 2000, *Service life prediction—State of the art report*, ACI Committee 365, Manual of Concrete Practice.

ACI, 2008, *Guide to durable concrete*, ACI Committee 201, ACI 201.2R-08.

Andrade, C., Polder, R., and Basheer, M., 2007, Non-destructive methods to measure ion migration, in *Non-Destructive Evaluation of the Penetrability and Thickness of the Concrete Cover—State-of-the-Art Report of RILEM Technical Committee 189-NEC*, R. Torrent and L. F. Luco (eds.), RILEM Publications SARL, Paris, France, Chapter 5.

Angst, U. M. and Elsener, B., 2014, On the applicability of the Wenner method for resistivity measurements of concrete, *ACI Materials Journal*, 111(6), 661–671.

Bentur, A., Diamond, S., and Berke, N., 1997, *Steel Corrosion in Concrete*, E&FN SPON/Chapman and Hall, London, UK.

Berke, N. S. and Hicks, M. C., 1990, Electrochemical methods of determining the corrosivity of steel in concrete, *ASTM STP 1000, 25th Anniversary Symposium Committee G1*, American Society of Testing and Materials, Philadelphia, PA, pp. 425–440.

Bickley, J., Hooton, R. D., and Hover, K. C., 2006, *Preparation of Performance-Based Specification for Cast-in-Place Concrete*, RMC Research Foundation.

Broomfield, J. and Millard, S., 2002, Measuring concrete resistivity to assess corrosion rates, Current Practice Sheet No. 128, *Concrete*, 37–39.

Carino, N. J., 1999, Nondestructive techniques to investigate corrosion status in concrete structures, *Journal of Performance of Constructed Facilities*, August, 96–106.

Carino, N. J., 2004, Methods to evaluate corrosion of reinforcement, in *Handbook on Nondestructive Testing of Concrete*, 2nd edition, V. M. Malhotra and N. J. Carino (eds.), CRC Press, Chapter 11.

Darts, 2004a, DARTS—Durable and reliable tunnel structures: Deterioration modelling, European Commission, Growths 2000, Contract G1RDCT-2000-00467, Project GrD1-25633.

Darts, 2004b, DARTS—Durable and reliable tunnel structures: Data, European Commission, Growths 2000, Contract G1RD-CT-2000-00467, Project GrD1-25633.

Duracrete, 2000, DuraCrete – Probabilistic Performance Based Durability Design of Concrete Structures, Final Technical Report, The European Union – Brite EuRam III, Project BE 95-1347, May 2000, information: COWI, Lyngby, Denmark.

Gowers, K. R. and Millard, S. G., 1999, Measurement of concrete resistivity for assessment of corrosion severity of steel using Wenner technique, *ACI Materials Journal*, 96(5), 536–541.

Hooton, R. D. and Emery, J. J., 1990, Sulfate resistance of a Canadian slag cement, *ACI Materials Journal*, 87(6), 547–555.

Kranc, S. C. and Sagues, A. A., 1992, Computation of corrosion macrocell current distribution and electrochemical impedance of reinforcing steel in concrete, in *Computer Modeling in Corrosion, ASTM STP 1154*, R.S. Munn (ed.), American Society of Testing and Materials, Philadelphia, PA, p. 95.

Lindgard, J., Andic-Cakir, O., Fernandes, I., Ronning, T. F., and Thomas, M. D. A., 2012, Alkali–silica reactions (ASR): Literature review on parameters influencing laboratory performance testing, *Cement and Concrete Research*, 42, 2012, 223–243.

Lobo, C. L., 2007, New perspective on concrete durability, *Concrete*, Spring, 24–30.

RILEM TC116, 1999a, RILEM Committee TC 116, Permeability of concrete as a criterion of its durability, *Materials and Structures*, 32, 174–179.

RILEM TC116, 1999b, RILEM TC116-PCD, Concrete durability—An approach towards performance testing, *Materials and Structures*, 32, 163–173.

Sagues, A. A., 1993, Corrosion measurement techniques for steel in concrete, Paper No. 353 in Corrosion 93, *The NACE Annual Conference and Corrosion Show*, NACE, Huston, TX.

Setzer, M. J., Auberg, R., Kasparek, S., Palecki, S., and Heine, P., 2001, CIF-test-capillary suction, internal damage and freeze thaw test: Reference method and alternative methods A and B, *Materials and Structures*, 34(9), 515–525.

Stearn, M. S. and Geary, A. J., 1957, Electrochemical polarization I: A theoretical analysis of the slope of the polarization curve, *Journal of the Electrochemical Society*, 104(1), 56–63.

Torrent, R. and Luco, L. F. (eds.), 2007, *Non-Destructive Evaluation of the Penetrability and Thickness of the Concrete Cover—State-of-the-Art Report of RILEM Technical Committee 189-NEC*, RILEM Publications SARL, Paris, France.

Wong, H. W. and Saraswathy, V., 2007, Corrosion monitoring of reinforced concrete structures—A review, *International Journal of Electrochemical Science*, 2, 1–28.

Chapter 9

Design of concrete mixtures for durability

9.1 INTRODUCTION

Concrete mix design refers to the process of

1. *Selecting* the suitable ingredients (cement, supplementary cementing materials [SCMs], aggregates, water, and admixtures) for a particular concrete mix
2. *Proportioning* these ingredients in such a way as to produce a concrete that in the *fresh state* can, without segregating, be mixed, transported, compacted, and finished, and in the *hardened state* achieves the required durability, strength, and dimensional stability

While a great deal of work has been done on the theoretical aspects of mix design (e.g., de Larrard, 1999), in practice it remains largely an empirical process, because of the large number of variables in the component materials, production, and exposure conditions of the concrete. Indeed, it is often considered to be more of an art than a science. In addition to this theoretical work, there are also many more books, manuals, guides, and technical papers dealing with the practical considerations involved in proper mix design (e.g., Shilstone, 1990; ACI Committee 211, 1991; Kosmatka et al., 2003; Day, 2006). As well, many concrete producers have their own "in-house" design procedures, tailored to their own sources of materials.

Before 1919, concrete mix design was a rather arbitrary procedure, with no specific guidelines on how to develop a suitable mix. For instance, Sabin (1905) merely states that "one should consider concrete...as a volume of aggregate bound together by a mortar of the proper strength. The volume of voids in the aggregate, the percent of the volume filled with mortar, and the strength of this mortar become then the important considerations in proportioning concrete." Proportioning was carried out by volume, rather than by weight (Thompson, 1909). Only after the publication of *Design of*

Concrete Mixtures by Abrams (1919) did concrete mix design start to be based on the w/c ratio "law."*

Abrams discussed concrete mix design almost entirely on the basis of strength and workability, and this remains largely true today. Only after certain concrete durability problems came to be understood (e.g., freeze–thaw cycles, sulfate attack, alkali–aggregate reactions [AARs]) were mix design procedures developed to deal with these issues. The first consideration in mix design even now still mostly remains compressive strength, with durability considered only secondarily. For instance, a recent book edited by Brandt (1998) makes no mention of durability, while de Larrard (1999) gives durability a scant three pages in a book of 423 pages. This is most unfortunate; most concrete problems that occur are durability-related, rather than due to inadequate strength. As well, the *sustainability* of concrete is also largely a function of its durability (e.g., Gjørv, 2009; Aïtcin and Mindess, 2011). Thus, concrete mix design procedures based first on durability considerations, and only secondarily on compressive strength, would seem to make more sense. However, design based upon durability is something that is still "more honor'd in the breach than the observance." Unfortunately, given the fixation of structural engineers on compressive strength to the exclusion of other concrete properties, this is unlikely to change anytime soon!

It is not the purpose here to provide a detailed description of concrete mix design procedures. These can be found in the references listed above, and from many other sources. Rather, the focus will be specifically on those aspects of mix design that concern the *durability* of the hardened concrete.

9.2 FUNDAMENTALS OF MIX DESIGN

For any concrete mix design method, the basic design considerations are the same:

- The *economy* of the mix, comprising the costs of the materials, labor, and equipment
- The *rheological properties* of the fresh mix, taking into account the equipment and personnel available
- The *compressive strength* of the hardened mix, including both the rate of early strength gain and the design strength
- The *durability* of the resulting hardened concrete, considering the expected environmental conditions to which the concrete will be exposed
- The specified *service life* of the structure

---

* Feret (1896) had formulated a more general expression of concrete strength in terms of the "cement/total voids" ratio, of which Abrams' law is a special case.

The mix design process thus requires a satisfactory resolution of *all* of these requirements. Since these often cannot all be optimized simultaneously, some compromises (as between strength and workability or durability and economy) may be required in any particular case. It should also be emphasized that even a "perfect" mix will not perform adequately if the proper mixing, placing, finishing, and curing procedures are not carried out.

Although the many different mix design procedures referred to above differ in detail, they all do in the end lead to very similar concrete mixes; the laws of physics and chemistry are, after all, universal. However, in any comparison of these different methods, it must be remembered that different jurisdictions may measure strength in quite different ways, so that a "North American" compressive strength determined from cylinder tests is quite different from (and generally smaller than) a "British" compressive strength determined by testing cubes. Moreover, cylinders (or cubes) of different sizes also will yield different strengths. Similarly, the different national test methods used to measure such things as sulfate concentrations may also lead to somewhat different values. It is therefore important, wherever possible, to work within the standards and specifications of the jurisdiction in which the structure is being built. Transporting one's own national standards into a different country is never advisable.

9.2.1 Design procedures

In general, most design procedures involve the following steps:

1. Information is required about the structural requirements, such as required strength, exposure conditions, types and dimensions of the structural members, reinforcing details, and any other special requirements. As well, the properties of the available coarse and fine aggregates such as their unit weights, sieve analyses, absorption capacities, and so on must be determined. Finally, the types, properties, and availabilities of the cementing materials (Portland cement, fly ash, silica fume, filler materials, and other pozzolans) should be known.
2. The required slump (or other measure of workability) should be specified; ideally, one would choose the lowest slump that will permit adequate placement and consolidation.
3. The maximum aggregate size should be selected. While the largest size allowable will minimize the required binder content, more commonly now rather smaller sizes are chosen to make placement of the fresh concrete easier.
4. Once the workability and maximum aggregate size are chosen, it is possible to estimate the amount of water (and entrained air if freeze-thaw exposure is a consideration) to provide the required workability.
5. It is now possible to estimate the amount of cement (or binder) required to provide a w/c (or w/b) that will provide the necessary strength and

durability. Most national standards have fairly detailed guidelines as to the selection of w/c, strength, and air content for different classes of structures, depending on the severity of the environmental exposure (see Chapter 5).

6. The volumes of coarse and fine aggregate are then determined, based on the fineness modulus of the sand and the maximum aggregate size and, of course, their specific gravities (or relative densities). Typically, the coarse and fine aggregate will occupy about 65%–70% of the concrete volume.

7. The relative amounts of the materials chosen may then be slightly adjusted to make sure that they make up a volume of exactly 1.00 m³.

8. Finally, and perhaps most importantly, trial batches of the selected mix design are prepared and tested, to ensure that all of the requirements of both the fresh and hardened concretes are met. If not, the mix is adjusted to correct any deficiencies.

Of course, it is not necessary to go through this procedure for each different project. The ready-mix concrete producers and large precast companies all have detailed charts that provide the mix designs for a wide variety of strength and durability requirements, including the use of blended cements. These are based on their experience with their locally available materials, and are generally highly reliable. Only for very special structures, or when having to use unfamiliar materials, are these detailed mix design procedures used.

The procedures outlined above should provide satisfactory concrete. However, it is worth emphasizing the special requirements for concretes exposed to severe exposure conditions. In what follows, a number of different specific durability cases are examined in more detail.

9.2.2 Freezing and thawing

Concrete exposed to repeated cycles of freezing and thawing when in a saturated or near-saturated (relative humidity >90%) state is susceptible to severe damage if appropriate measures are not taken. This problem is particularly severe in northern Europe, the northern parts of North America, and the northern parts of Asia. This is not a function of the chemistry of the Portland cement, but rather a function of the porosity and pore size distribution of the hardened cement paste. (The voids in the *concrete* resulting from incomplete compaction are most commonly filled with air rather than water, and so do not play a significant role in freeze–thaw damage.) It has long been known, however, that there is a relatively simple solution to this problem: incorporating a small volume of *entrained* air, to provide some empty space within the hardened cement paste in which the water can move and freeze without causing damage. This can easily be done using any of the many air-entraining admixtures commercially available. Typically, the spherical entrained air bubbles range in size from about 0.05 to 1.25 mm.

For good freeze–thaw resistance, the spacing factor (i.e., the average distance from any point in the paste to the edge of the nearest void) should be less than 0.2 mm. In general, the total air content in the concrete as measured by the standard tests should be in the range of about 5%–8%, depending upon the maximum aggregate size. Optimum frost resistance is achieved when the air content *in the cement paste* is about 17%. It should be noted, however, that the entrained air does have an adverse effect on strength; every additional 1% increase in entrained air reduces the compressive strength by between about 2% to 9%, depending on the cement source, admixtures, and the other mix ingredients (Kosmatka et al., 2003). This can be overcome by appropriately reducing the w/c ratio.

Like cement paste, aggregates themselves can be damaged by repeated cycles of freezing and thawing. The susceptibility of an aggregate to freeze–thaw damage depends on the porosity and permeability of the aggregate, the degree of saturation, its tensile strength, and the size of the particle. Fortunately, most aggregates have such a low porosity that the expansion of any water on freezing causes strains too small to initiate cracking. As well, even for complex pore systems, the critical degree of saturation is about 90% (Winslow, 1994). There is also a critical size below which freeze–thaw damage will not occur; for most aggregates, this is larger than the typical aggregate sizes used in construction. However, there are some aggregates (poorly consolidated sedimentary rocks such as some shales, cherts, or limestone) for which the critical size is less than the maximum size of aggregate used in practice. Such aggregates typically have a total absorption of water greater than 1.5%. Unfortunately, the various test methods that purport to evaluate the freeze–thaw resistance of aggregates tend not to be very reliable.

9.2.3 Sulfate attack

Sulfate attack is one of the most common and widespread forms of chemical attack on concrete. Sulfates are often present in soils and groundwaters, or may derive from industrial wastes such as mine tailings. If present in sufficient concentration, cracking and expansion of the concrete will occur, eventually leading to disintegration of the concrete, unless the concrete has been specifically designed to withstand sulfate attack. Sulfate attack is a highly complex process; depending on the particular sulfate, it may involve some or all of the cement hydration products. Though there is still considerable debate as to the precise mechanisms involved in sulfate attack, there is general agreement on the cause: the susceptibility of concrete to sulfate attack is directly related to the C_3A content of the cement; the higher the C_3A content, the more severe the attack. There are basically two mechanisms to be considered: formation of gypsum (*gypsum corrosion*) and formation of ettringite (*sulfoaluminate corrosion*).

Gypsum corrosion involves the reaction between sulfate ions and calcium hydroxide:

$$Ca(OH)_2 + SO_4^{2-}(aq) \leftrightarrow CaSO_4 \cdot 2H_2O + 2OH^-(aq) \qquad (9.1)$$

This reaction is accompanied by an expansion in solid volume of about 120%. It encourages the penetration of sulfate ions into the concrete and concentrates them in a form in which they can react directly with monosulfoaluminate.

Sulfoaluminate corrosion involves the formation of ettringite from monosulfoaluminate:

$$\underset{\text{monosulphate}}{Ca_4Al_2(OH)_{12} \cdot SO_4 \cdot 6H_2O} + \underset{\text{gypsum}}{2CaSO_4 \cdot 2H_2O} + 16H_2O$$
$$\rightarrow \underset{\text{ettringite}}{Ca_6Al_2(OH)_{12} \cdot (SO_4)_3 \cdot 26H_2O} \qquad (9.2)$$

This reaction is accompanied by a volume increase of about 55%; this volume expansion within the paste generates internal stresses that will eventually lead to cracking.

The most common sulfates are calcium sulfate, sodium sulfate, magnesium sulfate, and ammonium sulfate, and each reacts somewhat differently with hydrated Portland cement. As Taylor (1997) has pointed out, "the reactions of the cations and anions in…solution are essentially separate," which adds to the complexity of sulfate attack.

There is general agreement on how to mitigate sulfate attack: use cement with a low C_3A content and use a low w/c concrete. The severity of the attack dictates the details of how the concrete mix is to be designed. The Canadian requirements are given in Table 9.1.

Other national standards are essentially similar; they differ mostly in the limits on sulfate concentration used in their definitions of "severe" or "very severe." For higher sulfate concentrations, blended cements are most commonly used; the use of pozzolanic materials to replace some of the Portland cement effectively reduces the C_3A content of the binder.

Thaumasite sulfate attack is a particular form of sulfate attack that also involves the aggregates. Thaumasite is a complex calcium carbonate–silicate–sulfate hydrate ($CaCO_3 \cdot CaSiO_2 \cdot CaSO_4 \cdot 15H_2O$). The necessary conditions for its formation are a cold (<15°C), wet environment, sulfates or sulfides in the ground, moisture in the form of mobile groundwater, and carbonate-bearing aggregates such as limestones or dolomites. This form of attack ends with the conversion of the concrete to a mushy paste, as the C–S–H breaks down in the presence of the sulfate and carbonate ions and the binding properties of the matrix are destroyed.

Whether thaumasite is formed depends on the amount of $CaCO_3$ in the aggregate: the risk of attack is reduced when the carbonate fraction is below 10% of the fine aggregate or 30% of the coarse aggregate. As with all other forms of chemical attack, the severity of attack is reduced with high-quality, low w/c concrete.

Table 9.1 Requirements for concrete subjected to sulfate attack

Degree of exposure	Water-soluble sulfate (SO_4) in soil sample (%)	Sulfate (SO_4) in groundwater samples (mg/L)	Maximum water-to-cementing materials ratio	Minimum specified compressive strength (MPa) and age (days) at test	Air content category[a]	Cementing materials to be used
Very severe	>2.0	>10,000	0.40	35 within 56 days	1 or 2	High sulfate resistant
Severe	0.20–2.0	1500–10,000	0.45	32 within 56 days	1 or 2	High sulfate resistant
Moderate (including seawater exposure)	0.10–0.20	150–1500	0.50	30 within 56 days	1 or 2	Moderate sulfate resistant, or low heat of hydration

Source: Adapted from Canadian Standards Association, 2000, CSA Standard A23.1-00, Concrete Materials and Methods of Construction, Canadian Standards Association, Toronto, Canada.

Note: Table 9.1 is reproduced with the permission of Canadian Standards Association (operating as CSA Group), material is reproduced from CSA Group standards, A23.1-04/A23.2-04—Concrete Materials and Methods of Concrete Construction/Test Methods and Standard Practices for Concrete, which is copyrighted by CSA Group, 178 Rexdale Blvd., Toronto, ON, M9W 1R3. This material is not the complete and official position of CSA Group on the referenced subject, which is represented solely by the standard in its entirety. While use of the material has been authorized, CSA Group is not responsible for the manner in which the data is presented, nor for any interpretations thereof. No further reproduction or distribution permitted. For more information or to purchase standards from CSA Group, please visit http://shop.csa.ca/ or call 1-800-463-6727.

a   Air category 2 measured on fresh concrete requires air contents of 5%–8% for 10 mm maximum size of aggregate, 4%–7% for 14–20 mm aggregate, and 3%–6% for 28–40 mm aggregate; for air category 1, measured on hardened concrete, the corresponding values are 6%–9%, 5%–8%, and 4%–7%.

9.2.4 Alkali–aggregate reactions

The AAR (Blight and Alexander, 2011) refers to a reaction between alkalis (sodium and potassium) in the cement and certain types of aggregate. The main types of these reactions are the alkali–silica reaction (ASR), and the alkali–carbonate rock reaction (ACR).

The *ASR* involves the reactions between the alkalis in the pore solution and certain types of silica, such as volcanic glasses, opal, cristobalite, tridymite, chert, chalcedony, microcrystalline quartz, and so on. These reactions can lead to the formation of expansive alkali–silica gel in the concrete, which in turn may lead to extensive cracking. The *ACR*, on the other hand, does not produce a swelling gel. Rather, the coarse aggregate particles themselves expand, due to the reaction between the alkali hydroxides and small dolomite crystals within a clay matrix. This type of reaction is limited to carbonate aggregates containing clay, such as certain argillaceous dolomitic limestones. Fortunately, this type of reaction is quite rare, occurring primarily in certain parts of Canada. There is a substantial literature dealing with the mechanisms of ASR, which is summarized for instance in Blight and Alexander 2011, or in any of the standard concrete textbooks; it is thus not necessary to describe these mechanisms here. Instead, let us consider how to assess aggregates for potential alkali–aggregate reactivity, and how to mitigate against AAR if reactive aggregates must be used.

Ideally, of course, field performance over a long period of time is the best indicator of the potential reactivity of an aggregate. However, these data are often not available, and so a great many relatively short-term laboratory tests for reactivity have been developed, though they do not necessarily correlate all that well with field performance. As described by Alexander and Mindess (2005), these tests fall into three broad categories: nonquantitative tests used for initial screening; indicator tests to distinguish between reactive and nonreactive aggregates; and performance tests, providing information on how to avoid damaging expansions.

The most common *screening* test, and always a good place to start, is a detailed petrographic examination of an aggregate source by a qualified petrographer. The most common *indicator* tests involve producing mortar bars containing the suspect aggregate, and measuring the length change over a period of time, in some cases up to 1 year. Examples of such tests are ASTM C227 and ASTM C1260, which are mortar-bar tests for potential alkali–aggregate reactivity, and the RILEM TC 106 "ultra-accelerated mortar-bar test." *Performance* tests, such as ASTM C1293, employ larger prisms, so that actual concrete mixes may be tested; they also may be used to evaluate the effects of SCMs. These tests generally require a long duration, typically 1 year, for meaningful results to be obtained. Finally, long-term structural monitoring of actual structures must be used to monitor structures undergoing AAR, in order to assess their structural integrity over time.

The *ACR* most commonly involves reactions with very fine-grained dolomitic limestones [$CaMg(CO_3)_2$] containing some clay; these reactions too can cause expansion and cracking. Tests for these aggregates are described in ASTM C586, which involves immersing a small rock cylinder in an alkali solution, and ASTM C1105, which involves measuring the length change of a concrete bar.

9.2.4.1 Mitigation of AAR

Of course, the easiest and best way to guard against AAR damage is to use nonreactive aggregates, as indicated by the various tests mentioned above. If it does become necessary to use potentially reactive aggregates, however, there are a number of steps that can be taken to mitigate against AAR damage.

9.2.4.2 Use of a low alkali cement

It has been established experimentally (first by Stanton, 1940) that below a $\%Na_2O$ equivalent* of 0.6%, deleterious expansions due to ASR do not generally occur. (For ACR, the alkali content should be below 0.4% Na_2O equivalent.) More recent research (Hobbs, 1988) has suggested that the total alkali content of the concrete is also important; the upper limit on the Na_2O equivalent to avoid damage appears to be about 4 kg/m^3 (but may be significantly lower in some cases). Indeed, for concretes with high cement contents (such as many high-strength concretes), it has been suggested that a safe upper limit on alkalis would be about 3 kg/m^3. Unfortunately, over time, the alkali contents of cements have tended to increase, due to environmental regulations and changes in manufacturing technology.

9.2.4.3 Use of a low w/c

For AAR to occur, a supply of water is required to cause swelling of the alkali–silica gel. Therefore, low w/c concretes with a reduced permeability will at least delay the onset of deleterious expansions. If absolutely no external water is available, such expansions will not occur.

9.2.4.4 Use of SCMs

Fly ash, silica fume, metakaolin, ground-granulated blast-furnace slag, and other pozzolanic materials have been found to be effective in mitigating AAR, by effectively reducing the alkali content of the cementitious material;

---

* The $\%Na_2O$ equivalent is given by $\%Na_2O + 0.658\%K_2O$, where the factor 0.658 derives from the ratio of the atomic mass of Na_2O to K_2O. That is, the equation describes the *equivalent* effect of sodium in contributing alkalis.

typically, a 15%–20% cement replacement by Class F fly ash is sufficient. SCMs will also reduce the alkalinity of the pore solution, which is helpful. Silica fume can be particularly effective in reducing AAR, through both its pozzolanic action and its role in reducing the permeability of the concrete. This becomes particularly apparent in very high-strength concretes (Xincheng, 2013). There are several test methods to evaluate the efficacy of SCMs in reducing AAR expansion, such as ASTM C441 and ASTM C1567, both of which employ mortar-bar specimens.

9.2.4.5 Other chemical additions

Lithium compounds ($LiNO_3$ and $LiOH$) have also been found to reduce AAR expansions (Blight and Alexander, 2011), apparently due to the preferential formation of nonswelling lithium silicate hydrates.

9.2.5 Abrasion and erosion

The abrasion and wear of concrete are governed primarily by the aggregate, which is almost always harder and stronger than the cement paste (except for extremely high-strength concretes). This is a significant issue for concretes exposed to surface wear, such as pavements, airport runways, dam spillways, concrete canals carrying silt or gravel, or abrasion-resistant floor toppings. Abrasion of the aggregate in concrete will occur only after the aggregate particles become exposed due to the wearing away of the original concrete surface. In all cases, it is best to use aggregates that are hard, strong, and free of soft or friable particles. A low w/c ratio, high-strength concrete is also useful. It has also been reported (Bentur and Mindess, 2007) that the use of fiber reinforced concrete can improve the abrasion resistance. However, it must be remembered that stronger aggregates and a better paste–aggregate bond can only delay wear and abrasion, but cannot completely eliminate it (Mindess and Aïtcin, 2014).

The type of abrasive wear does, however, impose somewhat different aggregate requirements. For erosion due to suspended solids in flowing water, larger aggregate particles work better; for concrete subjected to cavitation, smaller (<20 mm) particles are preferred. For abrasion, the aggregate behavior becomes increasingly important as the concrete strength falls further and further below about 40 MPa.

9.2.6 Concrete exposed to sea water

Concrete in coastal areas will inevitably be exposed to sea water, which contains about 3.4% of dissolved salts. These salts are mostly chlorides and sulfates of sodium and magnesium (Table 9.2); all of these would be expected to adversely affect the properties of both fresh and hardened concrete.

Table 9.2 Typical composition[a] of sea water (ppm)

Sodium chloride (NaCl)	27,000
Magnesium chloride (MgCl$_2$)	3200
Magnesium sulfate (MgSO$_4$)	2200
Calcium sulfate (CaSO$_4$)	1100
Calcium chloride (CaCl$_2$)	500

[a] These are simply average values; they are slightly different in different seas and oceans, and can be quite diluted near the mouths of major rivers. For instance, values near the mouth of the Fraser River in British Columbia are about half of the numbers cited above. On the other hand, the Dead Sea has a salinity of about 34.2%!

9.2.6.1 Fresh concrete

Sea water should never be used as mixing water for concrete, except in the most extreme circumstances. The chlorides will lead to early corrosion of the steel reinforcement, and the sea water will lower the ultimate strength (though early strength gain will be accelerated because of the action of the chlorides).

9.2.6.2 Hardened concrete

One of the most severe environments to which hardened concrete may be exposed is seawater, particularly in structures subject to freezing and thawing cycles, such as the oil platforms in the North Sea. There are two different sets of mechanisms at play: deterioration of the concrete itself due to chemical attack by the seawater, exacerbated by freeze–thaw cycles and cycles of wetting and drying in the intertidal zone; and corrosion of the reinforcing steel by the chlorides in the water. There may also be abrasion of the concrete by ice floes.

Sea water is less corrosive than one might expect, given its chemical composition. This is in part due to the protective nature of the Mg(OH)$_2$, which tends to seal the concrete by filling in some of the pores. As well, gypsum and ettringite are more soluble in chloride solutions, which tend to limit deleterious expansions. For the concrete itself, the leaching of lime and calcium sulfate will lead to a gradual reduction in strength. The more serious problem is corrosion of the steel reinforcement.

There is no simple solution to these problems. Again, a low w/b, air-entrained concrete is a good start. For instance, for the Confederation Bridge in Eastern Canada (Aïtcin et al., 2016), located in a marine environment and subject to both freeze–thaw cycles and ice abrasion, it was ultimately decided to use a low alkali, blended cement containing 7.5% silica fume, an air content of 5% after pumping, and a w/c of ≈0.30. This provided a concrete with a compressive strength of 72 MPa that was able to meet the freeze–thaw requirements of Procedure A of ASTM C666 (300 cycles of freezing and thawing in water). The bubble spacing factor of the concrete was 350 μm. To help protect the steel, a concrete cover of 75 mm was specified, taking into account the abrasion of the concrete piers by moving ice. This example covers the elements needed for concrete designed to be used in such severe marine environments:

- Low w/b
- Cement blended with silica fume
- Low alkali cement
- Additional cover over the steel

A somewhat different approach has been used in the North Sea offshore structures built for the oil and gas industries. Those recommendations include a maximum w/c of 0.40 or less, a minimum cement content of 400 kg/m³, and sufficient air entrainment (Gjørv, 2009). A 2- to 3-mm-thick epoxy coating applied during slipforming was also found to be effective.

9.2.7 Acid attack

Hardened cement paste is an alkaline material; the pH of the pore water is about 13.0. Thus, concrete is subject to attack by acids in liquid form (concrete can be used to store *dry* chemicals).

The principal reaction in acid attack is the leaching of calcium hydroxide by hydrogen ions:

$$Ca(OH)_2 + 2H^+ \rightarrow Ca^{2+} + 2H_2O \tag{9.3}$$

If the acid is concentrated, there may also be attack on the C–S–H, forming silica gel:

$$\underset{\text{C–S–H}}{3CaO \cdot 2SiO_2 \cdot 3H_2O} + 6H^+ \rightarrow 3Ca^{2+} + \underset{\text{Silica gel}}{2(SiO_2 \cdot nH_2O)} + 6H_2O \tag{9.4}$$

Of course, as for all other forms of chemical attack, the concrete should be made as impermeable as possible. However, in this case, the chemistry of the binder should be considered: binders based on pozzolanic cements or on blast-furnace slag are more resistant to acid attack, because the CaO is combined in a less soluble form.

One particularly common form of acid attack is the attack by sulfuric acid in sewer pipes, where the sulfuric acid is produced through the action of certain anaerobic bacteria. These bacteria cause the formation of hydrogen sulfide; if the sewer is not running full, the hydrogen sulfide gas dissolves in the water films on the exposed concrete surfaces, leading eventually to the formation of sulfuric acid. This is considerably exacerbated by the action of other sulfide-oxidizing bacteria on the exposed pipe walls, which biogenically produce sulfuric acid. There are several measures that can be taken to increase the service life of concrete in such cases:

- Using limestone or dolomite aggregate instead of siliceous aggregate, since calcareous aggregates help to neutralize the acid.
- Chemical treatments of the concrete surface that precipitate insoluble salts to fill the pores, or to provide a resistant surface coating. Such chemicals include sodium silicate (water glass), fluoride salts, and certain iron compounds.
- The surface can be treated with carbon dioxide gas or silicon tetrafluoride vapor. These gases form highly insoluble $CaCO_3$ or CaF_2, respectively, which seal the surfaces.

However, it is probably more economical to design the pipes so that they always run full, though this is not easy in gravity sewers. A further measure is the use of calcium aluminate cements, which appear to have a stifling effect on biogenic acid production.

9.2.8 Corrosion of steel in concrete

The corrosion of steel in concrete is probably the most widespread and serious durability problem in North America and wherever chloride based deicing salts are used. As mentioned above, it is also a problem in marine structures. Corrosion of steel results in the formation of rust, which is an expansive reaction that can lead to cracking and spalling of the concrete. This is accompanied by a loss of steel cross section due to the corrosion, leading to a loss of load-bearing capacity of the structural member. Rusting is an electrochemical process that requires a flow of electrical current for the corrosion reactions to proceed. Oxygen and moisture must both be available for rusting to occur. The chemistry of steel corrosion in concrete has been described in detail elsewhere (Bentur et al., 1997; ACI 222R, 2001), as well as in Chapters 4 and 8 of this book, and will not be discussed here. The focus will be on how it can be prevented, or at least mitigated, in concrete design.

In order to protect steel against corrosion, it is necessary either to restrict the availability of water and oxygen at the steel surface, or to prevent electron flow within the steel. As stated repeatedly above, a low w/b concrete and the addition of some silica fume or other reactive cement extender will help to minimize the permeability of the concrete, and at least delay the penetration of chlorides to the level of the steel, and will also make it more difficult for water and oxygen to penetrate into the concrete. Protective membranes and specially designed overlays can also be helpful in this regard.

A number of *corrosion-inhibiting admixtures* are also now available. The most common one is calcium nitrite $(Ca(NO_2)_2)$, which is added to the mixing water. When used with high-quality concrete, and a sufficient depth of cover over the steel, it has been successful in suppressing the electrochemical process leading to corrosion. The calcium nitrite converts the passive film of ferrous oxide to γ-ferric oxide, which is then redeposited on the surface of the steel:

$$Fe^{2+} + (OH)^- + NO_2^- \rightarrow NO \uparrow + \gamma\text{-FeOOH} \tag{9.5}$$

There are a number of organic corrosion inhibitors also available commercially, primarily alkanolamines and amines and their salts (Elsener, 2011), combined with organic or inorganic salts. These are mostly proprietary formulations. Migrating corrosion inhibitors are also available. They are applied on the concrete surface, and then migrate through the concrete to the steel reinforcement.

At one time, fusion-bonded epoxy coatings on the steel surface were thought to be the most effective means of preventing steel corrosion. However, if the epoxy coating is damaged or scratched, very rapid localized corrosion will take place at these areas. With the advent of the corrosion inhibitors, the use of this "green" reinforcement is now less common. Similarly, stainless-steel reinforcing bars may be used. They are very effective in preventing corrosion, but are generally too expensive for "ordinary" use. They have been used successfully in extreme marine exposure conditions (Gjørv, 2009), where the increase in service life more than compensates for the high cost of the steel.

9.3 CONCLUDING REMARKS

In the design of concrete mixtures, durability considerations should be accorded the same importance as is given to strength and other mechanical properties. Possible destructive mechanisms should be identified at the beginning of a project, and it should then be possible to design a concrete mix that meets both strength and durability requirements. In practice, durability problems are more likely to occur through improper handling, placing, and curing of the concrete, or through poor design details such as inadequate drainage, than through poor materials selection. It should also be emphasized that to maintain adequate concrete behavior throughout the projected life of the structure, there must be a well-thought-out protocol for regular monitoring and maintenance of the structure, with any necessary repairs carried out in a timely manner.

REFERENCES

Abrams, D.A., 1919, *Design of Concrete Mixtures, Bulletin 1, Structural Materials Research Laboratory*, Lewis Institute, Chicago, IL, 20pp. (Reprinted from Minutes of the Annual Meeting of the Portland Cement Association, New York, December, 1918).

ACI Committee 211, 1991, *Standard Practice for Selecting Proportions for Normal, Heavyweight and Mass Concrete, ACI 211.1-91*, American Concrete Institute, Farmington Hills, MI.

ACI 222R, 2001, *Protection of Metals in Concrete against Corrosion*, American Concrete Institute, Farmington Hills, MI.

Aïtcin, P.-C. and Mindess, S., 2011, *Sustainability of Concrete*, Spon Press, Oxford, UK, 301pp.

Aïtcin, P.-C., Mindess, S., and Langley, W. S., 2016, The confederation bridge, in *Marine Concrete Structures*, M. G. Alexander (ed.), Elsevier, Cambridge, MA and Oxford, UK.

Alexander, M. and Mindess, S., 2005, *Aggregates in Concrete*, Taylor & Francis, Oxford, UK, 435pp.

ASTM C227, 2010, *Standard Test Method for Potential Aggregate Reactivity of Cement-Aggregate Combinations (Mortar Bar Method)*, American Society for Testing and Materials, West Conshohocken, PA.

ASTM C441, 2011, *Standard Test Method for Effectiveness of pozzolans or Ground Blast-Furnace Slag in Preventing Excessive Expansion of Concrete Due to the Alkali-Silica Reaction*, American Society for Testing and Materials, West Conshohocken, PA.

ASTM C586, 2011, *Standard Test method for Potential Alkali Reactivity of Carbonate Rocks as Concrete Aggregate (Rock-Cylinder Method)*, American Society for Testing and Materials, West Conshohocken, PA.

ASTM C666, 2008, *Standard Test Method for Resistance of Concrete to Rapid Freezing and Thawing*, American Society for Testing and Materials, West Conshohocken, PA.

ASTM C1105, 2008, *Standard Test Method for Length Change of Concrete Due to Alkali-Carbonate Rock Reaction*, American Society for Testing and Materials, West Conshohocken, PA.

ASTM C1260, 2007, *Standard Test Method for Potential Alkali Reactivity of Aggregate (Mortar Bar Method)*, American Society for Testing and Materials, West Conshohocken, PA.

ASTM C1293, 2008, *Standard Test Method for Determination of Length Change of Concrete Due to Alkali-Silica Reaction*, American Society for Testing and Materials, West Conshohocken, PA.

ASTM C1567, 2008, *Standard test method for Determining the Potential Alkali-Silica Reactivity of Combinations of Cementitious Materials and Aggregates (Accelerated Mortar-Bar Method)*, American Society for Testing and Materials, West Conshohocken, PA.

Bentur, A., Diamond, S., and Berke, N., 1997, *Steel Corrosion in Concrete: Fundamentals and Civil Engineering Practice*, E&FN Spon, London, UK.

Bentur, A. and Mindess, S., 2007, *Fibre Reinforced Cementitious Composites*, 2nd edition, Taylor & Francis, London, UK, 601pp.

Blight, G. E. and Alexander, M. G., 2011, *Alkali-Aggregate Reaction and Structural Damage to Concrete*, CRC Press/Balkema, Leiden, The Netherlands, 234pp.

Brandt, A. M., 1998, *Optimization Methods for Material Design of Cement-Based Composites*, E&FN Spon, London, UK, 314pp.

Canadian Standards Association, 2000, *CSA Standard A23.1-00, Concrete Materials and Methods of Construction*, Canadian Standards Association, Toronto, Canada.

Day, K. W. 2006, *Concrete Mix Design, Quality Control and Specification*, Taylor & Francis, Oxford, UK, 357pp.

de Larrard, F., 1999, *Concrete Mixture Proportioning*, E&FN Spon, London, UK, 421pp.

Elsener, B., 2011, Corrosion Inhibitors—An Update on On-Going Discussion, 3, Länder Korrosianstagung "Möglichkeiten des Korrosionsschutzes in Beton" 5/6. Mai. Wien GfKorr, Frankfurt am Main, German, pp. 30–41.

Feret, R., 1896, Essais de Divers Sables et Mortiers Hydrauliques, *Annales des Ponts et Chaussées Mémoires et Documents, Série 7*, Tome VII, deuxième semester, pp. 174–197.

Gjørv, O. E., 2009, *Durability Design of Concrete Structures in Severe Environments*, Taylor & Francis, Oxford, UK, 220pp.

Hobbs, D. W., 1988, *Alkali-Silica Reaction in Concrete*, Thomas Telford, London, UK, 183pp.

Kosmatka, S. H., Kerkhoff, B., and Panarese, W. C., 2003, *Design and Control of Concrete Mixtures*, 14th edition, Portland Cement Association, Skokie, IL, 358pp.

Mindess, S. and Aïtcin, P.-C., 2014, How can we move from prescription to performance? in *Proceedings of the RILEM International Workshop on Performance-Based Specification and Control of Concrete Durability*, H. Bjegovic, H. Beushausen and M. Serdar (eds.), RILEM Proceedings PRO 89, RILEM Publications, Bagneux, Paris, pp. 267–273.

RILEM TC-106, 2000, Recommendations of RILEM TC 106-AAR: Alkali-aggregate reaction, *Materials and Structures*, 33(229), 283–293.

Sabin, L. C., 1905, *Cement and Concrete*, Archibald Constable and Co., London, UK, 507pp.

Shilstone, J. M., Sr., 1990, Concrete mixture optimization, *Concrete International*, 12(6), 33–39.

Stanton, T. E., 1940, Expansion of concrete through reaction between cement and aggregate. *Proceedings ASCE*, 66, 1781–1811.

Taylor, H. F. W., 1997, *Cement Chemistry*, 2nd edition, Thomas Telford, London, UK.

Thompson, S. E., 1909, *Concrete in Railroad Construction*, Atlas Portland Cement Company, New York, NY, 228pp. (Reprint published by the National Model Railroad Association, Chattanooga, TN, 2010.)

Winslow, D., 1994, The pore system of coarse aggregates, in *Significance of Tests and Properties of Concrete and Concrete Making Materials*, ASTM STP 169C, American Society for Testing and Materials, West Conshohocken, PA, pp. 429–427.

Xincheng, P., 2013, *Super-High-Strength High Performance Concrete*, CRC Press, Boca Raton, FL, 258pp.

Chapter 10

Durability and construction

10.1 INTRODUCTION

Other chapters in this book have dealt with, *inter alia*, durability modeling and design, specifications, use of durability indicators, durability testing, and concrete mixtures for durability. These all relate to efforts by designers and owners of infrastructure to achieve durable concrete construction. Ultimately, durability is achieved, or frequently *not* achieved, during actual construction. The best design intentions and use of the most appropriate materials will be in vain if these intentions and the materials' potential for durability are not realized during construction.

The achievement of durability during the construction process is not trivial. The many and varied operations on a construction site, whether for a relatively small building or a large complex infrastructure project, all contribute in different ways to the final quality of the built structure. Thus, every step in the construction process needs to be carefully planned and controlled by the constructor to achieve the desired quality of construction.

Many practical construction details are beyond the scope of this chapter, and other comprehensive texts should be consulted (e.g., ACI 304R-00; Murdock et al., 1991; Owens, 2009*; Kosmatka et al., 2003). This chapter will cover certain critical issues in the practical achievement of durability by focusing on aspects such as concrete manufacture, construction challenges and site practices, and quality control for durability. Two brief case studies of durable construction are also given at the end.

The terms "durability" and "quality" need definition. "Durability" has several formal definitions (see Chapter 3) relating to the ability of the structure to perform its intended function in the design environment over its specified life. The emphasis is on functionality and there is acceptance of a notional service life. "Quality" is a broader and more complex term. According to the Oxford Dictionary, it is *"The standard of something as measured against other things of a similar kind; the degree of excellence*

---

* Chapter 12 of Owens (2009)—*Fulton's Concrete Technology* (9th edition) was useful in the preparation of this chapter, and this source is hereby acknowledged.

of something." Thus, quality has to do with a standard of measurement; we will take this to be the requirements of the construction specification, although as noted previously, these can themselves be ambiguous and vague. This requires dialogue between all concerned parties (client, engineer, constructor, and material supplier) at the commencement of the work to produce a durable structure. It also underlines the need for unambiguous and clear performance-based specifications as a necessary though not sufficient prerequisite for improving the quality of concrete construction. In this chapter, "durability" and "quality" will be taken to refer broadly to the same idea, that is, conformity with a given standard of performance.

10.2 CONCRETE MANUFACTURE

Concrete manufacture and construction are multistep processes, illustrated in Figure 10.1. They involve materials selection (depending on the type of specification) and sampling, batching and mixing, and final finishing and protection afforded the freshly cast concrete structure. Figure 10.1 relates mostly to the activities of the concrete constructor, where experience and expertise are essential for high-quality construction. The skills involved in good concrete construction are as demanding as in any other industry and need serious attention. Producing high-quality concrete is not particularly difficult, provided attention is paid to the processes indicated in Figure 10.1. What *is* challenging is to produce high-quality concrete *consistently*.

This section concentrates mainly on batching and mixing of concrete—whether on-site or in a ready-mix operation. Other processes in Figure 10.1 are equally important, however. Advantages and disadvantages attach to whether to use site production or ready-mix production. In the end, the quality of the concrete as delivered to the final point of placing or discharge is the critical issue; it is possible to obtain quality concrete from either production process.

10.2.1 Batching of concrete mixes

Batching is the process of measuring the concrete ingredients accurately and consistently, to produce mixes that are consistent in quality and in properties related to processing or placing, such as workability. Batching may be carried out by mass or by volume. The latter is not favored for production of high-quality concrete mixes because of difficulties in obtaining consistent proportions in the mix; for example, both sand and cement may have very different loose bulk densities depending on their moisture and aerated state.

Batching systems and equipment should be chosen to suit the particular operation. Essential elements of such systems are reliable means of moving

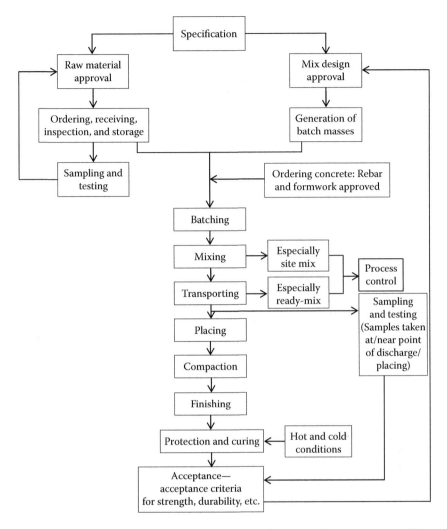

Figure 10.1 Processes involved in concrete manufacture and construction. (After Owens, G. (ed.), 2009, *Fulton's Concrete Technology*, 9th edition, Cement & Concrete Institute, Midrand, South Africa, Figure 12.1, reproduced with permission from the Concrete Institute.)

materials around the plant; robust materials storage facilities, for example, silos or bins; accurate and reliable mass-measuring devices that are regularly calibrated and checked; and suitable protection from the elements and environment so that ingredients are not contaminated. Daily checking and regular calibration of mass-measuring devices is very important. Guidance can be found in, for example, BS EN 206-1, ASTM C94, and CSA A23.1/A23.2.

Batching operations may be manual, semiautomatic, or fully automatic. Automated systems are needed for large production volumes where monitoring of material flows is continuous, with electronic switches being activated to cut off flow when the required mass is reached. Interlocks are included to interrupt an automatic cycle when preset mass tolerances are exceeded. Automated systems should record the mix details and also permit stock control since quantities of batched materials are continuously recorded. Guidance can be found in, for example, SANS 878 and ASTM C94.

10.2.1.1 Control of mixing water content

Control of mixing water content during the batching process is probably the single most important operation (other than control of binder or cement content) that influences consistent quality of concrete. In practice, final control of water content is often done by observing the consistence (slump or flowability) of concrete in the mixer or truck and withholding or adding increments of water accordingly. This results in variable water contents in the delivered mixes. Together with the poor practice of uncontrolled addition of water at the final point of discharge if the slump appears too low (this is often not assessed properly), the consequence is that on-site concrete may contain highly variable water contents that can have serious implications not only for durability but also for concrete strength.

ASTM C94/C94M-14b and ACI 117-10 state that the total mixing water should be within ±3% of the design mix quantity. Total mixing water includes all water added to the batch during mixing, water from aggregate moisture content, admixture water content, and so on. Most chemical admixtures contain in excess of 60% water and this should be included in the mixing water. Regarding batch mixing water, ASTM C94 requires that this be accurate to within ±1% of the design quantity. On the basis of a mix water requirement of, say, 200 L/m³, the above tolerances translate into ranges of 12 and 4 L/m³, respectively. This could amount to a difference of 0.04 w/c ratio units (which could double if the tolerance for cement batching is added), which is very significant, with consequences for variation of *in situ* properties, including durability.

The added water content is strongly influenced by the water content of the aggregates, and this makes consistent concrete water content a challenge. Water contributed by aggregates can be substantial, up to 50 L/m³ of total water content or more. Therefore, errors in estimating aggregate moisture content have large influences on mixing water content and therefore also on w/c ratio and consistence. Aggregate moisture content should be measured or estimated to within 1% error, preferably by using laboratory methods or calibrated aggregate moisture meters, which allow automatic compensation of batch water. Water content can then be controlled by a combination of adjustment due to moisture in the aggregate particularly the sand, observation of consistence in the mixer, monitoring of mixer torque, and carrying out of regular slump or other suitable workability tests.

Water is also added to ready-mix trucks at various points, often determined by the truck operator based on visual observation. ASTM C94 allows a one-time addition of water so that concrete slump can be brought to specified levels, provided the total design mixing water content is not exceeded.

A further factor is the influence of temperature and delivery time on mixing water requirement. An NRMCA* study (Gaynor et al., 1985) showed that a mix designed for a concrete temperature of 18°C and delivery time of 20 minutes could have an increase in mixing water requirement of approximately 40 L/m³ if the mix temperature increased to 35°C and delivery time to 90 minutes. Such situations will obviously lead to increased standard deviations in strength, and more importantly to durability variations. (Refer to Gaynor et al. [1985], for further details.)

10.2.1.2 Quality control of concrete constituents and concrete production

To produce good quality concrete consistently requires that the quality and consistency of the constituent materials are maintained during the job, and that close attention is given to all aspects of concrete production. A sampling and testing program for quality control of concrete materials and production must be introduced. BS EN 206-1 has an entire clause (Cl. 9) dedicated to the issue of production quality control. According to this standard, production control comprises all measures necessary to maintain the properties of concrete in conformity to specified requirements, including

- Production control systems
- Recorded data and other documents
- Testing
- Concrete composition and initial testing
- Personnel, equipment, and installation
- Batching of constituents
- Mixing of concrete
- Production control procedures

The relevant sections of BS EN 206-1 take account of the principles of EN ISO 9001 (which is concerned with the criteria for a quality management system).

Some practical issues that should be considered are mentioned here.

1. Where possible, raw materials should be inspected in the trucks before unloading. Aggregates should be inspected for cleanliness, size, color, and contamination. Admixture containers must be clearly identified

---

* National Ready Mixed Concrete Association (USA).

and stored such that the contents are not exposed to extreme temperatures. The seals on the inlet and discharge valves of tankers for cementitious materials must be intact, must be the correct color, and carry the correct code/serial/delivery number. Care must be taken to discharge into the appropriate silo.

2. Materials must be handled and stored in such a way that their quality does not deteriorate. Aggregates may be stored in separate stockpiles, ground storage bays, or overhead bins. Tipping and handling of aggregates to prevent segregation and intermingling are important. Stockpiles placed close together require partitions to prevent contamination or mixing with adjacent materials. Suitable drainage is also required, for example, a concrete floor laid with a fall. When loading aggregates from stockpiles on the ground, care must be taken by the loader operator not to contaminate the aggregate by scraping too low. This not only contaminates the aggregate but also causes a depression in which water will accumulate. The operator should be instructed to leave at least 300 mm of aggregate on the ground. Where possible, during inclement weather, variations in the aggregate moisture content can be minimized by covering with tarpaulins or heavy plastic sheeting. During hot weather, coarse aggregates may be cooled by water spraying (with suitable allowance made for adjustment to mixing water content), while in cold areas, frosting of aggregates should be minimized by steam heating or other methods of raising the temperature or preventing freezing.

3. Cement may be delivered in bulk by road or rail tanker, or in bags. Bagged cement requires special precautions during storage, to prevent contamination and exposure to moist or humid conditions. A clear system of "first in, first out" is required to ensure that cement is used in the order in which it is delivered. Obviously, if cement silos are used for storage, which is the preferred method, the cementitious materials are better protected and used in the correct order of delivery.

Quality control for durability is specifically covered later in this chapter.

10.2.2 Concrete mixing

Concrete mixing is the process of ensuring that all ingredients are intimately combined and evenly distributed throughout the mass of fresh concrete, which is essential for homogenous and uniform concrete on the job site. Mixing is an advanced technology in its own right, and specialist texts should be consulted, for example, on types and proper operation of mixers, machine versus hand mixing, and so on (see, e.g., ASTM C94M-14b). Machine mixing is preferred, if not mandatory, for high-quality jobs because of the better uniformity of mixing achieved. Mixing can be accomplished in continuous or batch mixers of various designs. Pan mixers,

Figure 10.2 Inside of modern mobile pan mixer.

such as the one shown in Figure 10.2, while sometimes of lesser capacity than other types of mixers, are forced action mixers that can thoroughly mix even low-slump concretes with mixing times of less than 1 minute per batch. They are thus very efficient.

Truck mixers are usually of the reversing drum type. Ready-mix concrete plants may have integral concrete mixers in their plant, or they may rely on the truck to effect the mixing. Further agitation of the mix may occur *en route* to the job site. In this case, adequate time must be allowed for the concrete to be thoroughly mixed before discharge. The efficiency of truck mixers is dependent upon cleanliness of blades, revolutions per minute, and whether the blades are designed for mixing or agitation.

10.3 SITE CONSTRUCTION: CHALLENGES AND SITE PRACTICES

Achieving durable concrete construction involves producing concrete to the correct specification, mixing it to uniform consistency, and then transporting, placing, and compacting it into the formwork with subsequent finishing and curing such that the inherent properties of the concrete, including its potential durability, are not compromised. This involves a series of steps, each of which may result in the concrete not achieving its potential properties in the actual structure. Proper attention to detail in the processes of site concreting is absolutely essential to achieving quality concrete.

As indicated earlier, specialist texts or commercial advice should be consulted for details on site concreting practices. However, there are particular practices or processes where the possibility of the concrete quality being compromised is high, and these are discussed below.

10.3.1 Concrete handling, placing, and compaction

A basic requirement of handling and placing is to maintain the potential quality and uniformity of the concrete. Because placing and compaction are done almost simultaneously and are interdependent, they should be considered as one operation. Placing equipment should be capable of delivering the concrete as close as possible to its final position without segregation. Segregation may occur particularly during discharge, when concrete is allowed to free-fall or drop continuously and collect in a heap, or when discharge is from the ends of chutes or conveyors. If this occurs, it is very difficult to re-mix the concrete to a consistent and uniform mass. Figure 10.3 indicates how to control discharge of concrete to avoid harmful segregation (ACI 304-R-00). On large projects like dams, where larger aggregate sizes are used and placement by conveyors is justified, segregation without baffles or drop chutes is likely, mainly because of the larger (perhaps gap-graded) aggregates, lower workability, and speed at which the concrete is transported/placed in large-volume concrete pours. Table 10.1 provides a summary of the various methods and equipment for transporting and handling concrete.

Once the concrete has been correctly placed in its final position, it must be thoroughly compacted to expel entrapped air so as to achieve maximum density, strength, and durability—assuming good curing is carried out. Air voids include both entrapped air and voids from pockets or lenses of excess water. As a rule of thumb, every additional 1% increase in entrained air reduces the compressive strength by between about 2% to 9%, depending on the cement source, admixtures, and the other mix ingredients (Kosmatka et al., 2003), and the effects on penetrability (and hence potential durability) are likely to be equally if not more serious.

Proper compaction requires consideration of the placing method and concrete workability in relation to the compaction method chosen, and also appropriate layer thicknesses of the concrete so that full compaction can be done. Subsequent consolidation will not rectify or improve poorly placed concrete which has segregated or has been placed in thick layers. This is particularly important in walls, columns, or other thin vertical elements. In general, mechanical compaction (poker, screed, or shutter vibrators) is preferred to hand compaction, but any method that ensures that the air is thoroughly excluded from the concrete and that the final product represents as uniform and homogenous a material as possible is acceptable. It is also possible to segregate concrete during consolidation, particularly with higher slump concretes that lack cohesiveness, by causing the heavier aggregates to settle and allowing water and "slush" to be displaced upwards. Care must be taken with the choice of appropriate mechanical vibration systems

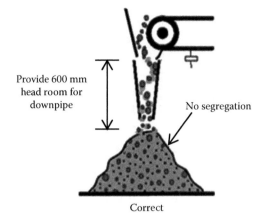

Provide 600 mm
head room for
downpipe

No segregation

Correct

The arrangement prevents segregation
of concrete whether it is being discharged
into vehicles or forms

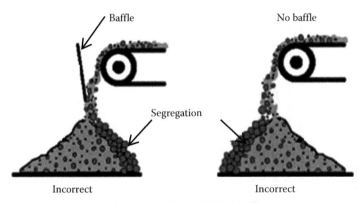

Baffle

No baffle

Segregation

Incorrect

Incorrect

Lack of control at the end of belt; a baffle can
change the direction of segregation. The above
occurs particularly with large aggregate size concrete

Figure 10.3 Avoidance of segregation during discharge of concrete into heaps. (Authorized reproduction from ACI 304 R-00, 2009, *Guide for Measuring, Mixing, Transporting, and Placing Concrete*, American Concrete Institute, Farmington Hills, MI.)

in relation to the type of concrete being compacted (e.g., low slump). Labor also needs to be well trained in the purpose and use of the equipment.

With the increasing use of self-compacting concrete (SCC), the understanding, skills, and art of good placing techniques may seem less important. However, while compacting may seem less necessary with SCC, the vast majority of concrete is still placed and compacted by conventional means, and this is likely to continue into the foreseeable future.

Table 10.1 Methods and equipment for transporting and handling concrete

Equipment	Type and range of work for which equipment is best suited	Advantages	Points to watch for
Belt conveyors	For conveying concrete horizontally or to a higher or lower level. Usually positioned between main discharge point and secondary discharge point	Belt conveyors have adjustable reach, traveling diverter, and variable speed both forward and reverse. Can place large volumes of concrete quickly when access is limited	End-discharge arrangements needed to prevent segregation and leave no mortar on return belt. In adverse weather (hot, windy), long reaches of belt need cover
Belt conveyors mounted on truck mixers	For conveying concrete to a lower, horizontal, or higher level	Conveying equipment arrives with the concrete. Adjustable reach and variable speed	End-discharge arrangements needed to prevent segregation and leave no mortar on return belt
Buckets	Used with cranes, cableways, and helicopters for construction of buildings and dams. Convey concrete directly from central discharge point to formwork or to secondary discharge point	Enables full versatility of cranes, cableways, and helicopters to be exploited. Clean discharge. Wide range of capabilities	Select bucket capacity to conform to size of the concrete batch and capacity of placing equipment. Discharge should be controllable. Vertical discharge results in less segregation than banana buckets
Chutes on truck mixers	For conveying concrete to a lower level, usually below ground level, on all types of concrete construction	Low cost and easy to maneuver. No power required; gravity does most of the work	Slopes should range between 1:2 and 1:3 and chutes must be adequately supported in all positions. End-discharge arrangements (downpipe) needed to prevent segregation

(Continued)

Table 10.1 (Continued) Methods and equipment for transporting and handling concrete

Equipment	Type and range of work for which equipment is best suited	Advantages	Points to watch for
Cranes and buckets	The right equipment for work above ground level	Can handle concrete, reinforcing steel, formwork, and sundry items in bridges and concrete-framed buildings	Has only one hook. Careful scheduling between trades and operations is needed to keep crane busy
Dropchutes (also tremie)	Used for placing concrete in vertical forms of all kinds. Some chutes are one piece tubes made of flexible rubberized canvas or plastic, others are assembled from articulated metal cylinders (elephant trunks)	Dropchutes direct concrete into formwork and carry it to bottom of forms without segregation. Their use avoids spillage of grout and concrete on reinforcing steel and form sides, which is harmful when off-the-form surfaces are specified. They also will prevent segregation of coarse particles	Dropchutes should have sufficiently large, splayed-top openings into which concrete can be discharged without spillage. The cross section of dropchute should be chosen to permit inserting into the formwork without interfering with reinforcing steel
Mobile batcher mixers	Used for intermittent production of concrete at jobsite, or where only small quantities are required	A combined materials transporter and mobile batching and mixing system for quick, precise proportioning of specified concrete. One-person operation	Trouble-free operation requires good preventive maintenance program on equipment. Materials must be identical to those in original mix design
Nonagitating trucks	Used to transport concrete on short hauls over smooth roadways	Capital cost of nonagitating equipment is lower than that of truck agitators or mixers	Concrete slump should be limited. Possibility of segregation. Height is needed for high lift of truck body upon discharge
Pneumatic guns (shotcrete)	Used where concrete is to be placed in difficult locations and where thin sections and large areas are needed	Ideal for placing concrete in freeform shapes, for repairing structures, for protective coatings, thin linings, and building walls with one-sided forms	Quality of work depends on skill of those using equipment. Only experienced nozzlemen should be employed

(Continued)

Table 10.1 (Continued) Methods and equipment for transporting and handling concrete

Equipment	Type and range of work for which equipment is best suited	Advantages	Points to watch for
Pumps	Used to convey concrete directly from central discharge point at jobsite to formwork or to secondary discharge point	Pipelines take up little space and can be readily extended. Delivers concrete in continuous stream. Pump can move concrete both vertically and horizontally. Truck-mounted pumps can be delivered when necessary to small or large projects. Tower-crane mounted pump booms provide continuous concrete for tall building construction	Constant supply of freshly mixed concrete is needed with average consistency and without any tendency to segregate. Care must be taken in operating pipeline to ensure an even flow and to clean out at conclusion of each operation. Pumping vertically, around bends, and through flexible hose will considerably reduce the maximum pumping distance
Screw spreaders	Used for spreading concrete over large flat areas, such as in pavements and bridge decks	With a screw spreader, a batch of concrete discharged from a bucket or truck can be quickly spread over a wide area to a uniform depth. The spread concrete has good uniformity of compaction before vibration is used for final compaction	Screw spreaders are normally used as part of a paving train. They should be used for spreading before vibration is applied
Tremies (also dropchutes)	For placing concrete underwater	Can be used to funnel concrete down through the water into the foundation or other part of the structure being cast	Precautions are needed to ensure that the tremie discharge end is always buried in fresh concrete, so that a seal is preserved between water and concrete mass. Diameter should be 250–300 mm (10–12 in.) unless pressure is available. Concrete mixture needs more cement, 390 kg/m^3 (658 lb/yd^3), and greater slump, 150–230 mm (6–9 in.), because concrete must flow and consolidate without any vibration

(Continued)

Table 10.1 (Continued) Methods and equipment for transporting and handling concrete

Equipment	Type and range of work for which equipment is best suited	Advantages	Points to watch for
Truck agitators	Used to transport concrete for all uses in pavements, structures, and buildings. Haul distances must allow discharge of concrete within 1.5 hours, but limit may be waived under certain circumstances	Truck agitators usually operate from central mixing plants where quality concrete is produced under controlled conditions. Discharge from agitators is well controlled. There is uniformity and homogeneity of concrete on discharge	Timing of deliveries should suit job organization. Concrete crew and equipment must be ready on-site to handle concrete
Truck mixers	Used to transport concrete for uses in pavements, structures, and buildings. Haul distances must allow discharge of concrete within 1.5 hours, but limit may be waived under certain circumstances	No central mixing plant needed, only a batching plant, since concrete is completely mixed in truck mixer. Discharge is same as for truck agitator	Timing of deliveries should suit job organization. Concrete crew and equipment must be ready on-site to handle concrete. Control of concrete quality is not as good as with central mixing
Wheelbarrows and buggies	For short flat hauls on all types of onsite concrete construction, especially where accessibility to work area is restricted	Very versatile and therefore ideal inside and on jobsites where placing conditions are constantly changing	Slow and labor intensive

Source: Reproduced from Kosmatka, S. H., Kerkhoff, B., and Panarese, W. C., 2003, *Design and Control of Concrete Mixtures*, *EB001*, 14th edition, Portland Cement Association, Skokie, IL, with permission from PCA.

Figure 10.4 Detail of fabric used for controlled permeability form liner, to allow near-surface drainage of excess water and expulsion of air from cast face of concrete. The filter layer is made of thermally bonded polypropylene fibers laminated to a plastic mesh with openings of 4 mm. (Courtesy of G. Evans.)

Re-vibration, done typically 1–2 hours after placing the concrete, is a technique that, if correctly applied, can assist in improving *in situ* quality of the concrete by eliminating defects from bleeding, plastic shrinkage, and restrained settlement. It is particularly effective in concrete displaying prolonged bleeding under evaporative conditions and settlement of the solids. Re-vibration is thus not harmful provided the concrete is still plastic enough to respond to vibration, and the hole left by the poker vibrator can still be closed. Unfortunately, there is little data showing the effects of re-vibration on, say, penetrability of concrete, but the effects are likely to be positive.

If a high-quality and impermeable cover layer of concrete is required, controlled permeability formwork (CPF) may be used. This is a cellular synthetic liner that is fixed to the inside of the forms and permits drainage of excess water from the surface after consolidation, while trapping the fine materials such as cement and fine aggregate particles. Thus, it ensures that the surface layer is not adversely affected by a higher w/b ratio than the bulk of the concrete and also assists in densifying this layer. Figure 10.4 shows a typical CPF fabric.

10.3.2 Protection and curing

Protection is normally applied to concrete in its fresh state (before it sets), and curing is normally applied to concrete after setting or removal of formwork. Both protection and curing are processes to maintain the humidity and temperature conditions to ensure that the potential quality of concrete is achieved through effective hydration.

Freshly placed concrete can deteriorate rapidly if it is not adequately protected immediately after placement to mitigate the adverse effects of early drying, high temperature, impact from rain, and so on. Since the external "skin" of the concrete is most severely affected by a lack of early protection, the consequent effects on durability particularly of reinforced concrete are obvious. Unfortunately, early protection and curing are often neglected with serious loss of potential durability of the concrete. It is not an exaggeration to state that poor protection and curing practices are second possibly only to inadequate steel cover in being the major causes of poor concrete durability.

Early protection is especially critical where rapid loss of moisture can occur from the concrete surface, due to environmental conditions of low relative humidity, high wind speed, and high ambient and concrete temperatures. Particularly at risk are large expanses of flatwork such as concrete floor slabs. In addition to adversely affecting the early hydration reactions and the capillary pore structure, loss of moisture causes shrinkage that may induce plastic cracking, which will permanently impair the concrete quality. Under evaporative conditions, highly cohesive and low bleed concretes such as SCC and those containing CSF or fine fillers are very susceptible to these effects. Suitable protection measures are erection of windbreaks and sunshades for slabs placed in exposed conditions; placing of moisture-retaining coverings during delays between compacting and finishing; judicial use of fog sprays above the surface to maintain a water sheen on the concrete before final finishing and the start of curing; carrying out concreting operations during the cooler part of the day; and the use of evaporation retardants (spray-applied aliphatic alcohol-based films).

Curing is conceptually simple: provision of suitable moisture and temperature conditions in the early period of the concrete's life for the cement to hydrate as fully as possible. In practice, however, curing is often difficult to carry out effectively, depending also on the type of element, for example, slabs, columns, walls, and so on. In general, any surface of a concrete element that is unprotected and exposed to the environment during the first days or weeks will result in a deterioration of properties of the surface layer, which has major implications for durability. The effects on strength of the element are likely to be less serious, since the interior of a member is protected by the concrete cover and tends to cure reasonably well. Aspects of concrete curing are covered below with an emphasis on concrete durability. (Refer to Chapter 6 [section on durability indicators or indexes] where the importance of measuring the surface or cover properties of concrete was stressed, particularly in respect of potential durability.)

Curing allows mixing water—essential for ongoing hydration—to be retained in the concrete. Where w/c is lower than about 0.5, curing should also provide additional water for hydration. Thus, curing practices are aimed at either moisture retention or moisture addition (or both). Delay in applying active curing will result in some drying of the surface, with the resultant deposition of calcium hydroxide in the near-surface capillaries, which tends to block them and hinders efforts at subsequent water ingress. Curing by intermittent wetting is similarly ineffectual. Curing needs to commence as soon as practicable and be applied continuously, as even a short delay or a break of continuity in hot dry weather can irreparably damage the concrete cover layer. Figure 10.5 shows the effect of concrete and air temperatures, relative humidity, and wind velocity on evaporation of surface moisture from concrete. As mentioned, curing is most critical to preserve and develop the properties of the cover layer, which has to resist chemical and physical attack and which must be as impermeable as possible. Proper curing of the cover layer helps to reduce the rate of ingress of contaminants such as chlorides and carbon dioxide, protecting the steel reinforcement from corrosion.

Blended cements with FA, GGBS, and CSF react more slowly than plain Portland cement, particularly in cold weather, and require additional protection and extended curing. Low temperatures slow the rate of hydration and require longer curing times, while high temperatures accelerate the moisture loss at early ages, which will have detrimental effects on long-term

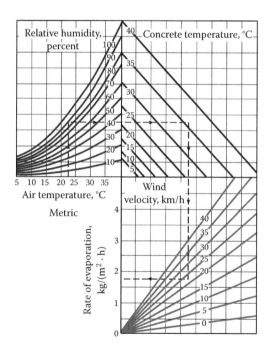

Figure 10.5 Chart to estimate the evaporation of water from freshly placed concrete. (Reproduced from Kosmatka, S. H., Kerkhoff, B., and Panarese, W. C., 2003, *Design and Control of Concrete Mixtures*, EB001, 14th edition, Portland Cement Association, Skokie, IL, with permission from PCA.)

strength and durability. Curing temperatures and evaporative conditions should be controlled to avoid these negative effects.

Curing practices are summarized in Table 10.2, while Table 10.3 gives guidance on the duration of moist curing required (from the South African Concrete Code). In this respect, note that the period of curing should ideally be prescribed in terms of developing the required properties of a concrete to the required extent. Regarding Table 10.3, the SA concrete code

Table 10.2 Guide to site curing practices

Type of construction	Curing material	Application
Road and airfield pavements	Pigmented resin-based curing compound with high efficiency rating	Immediately after finishing process is complete (concrete may have to be protected from rapid drying out before it sets to prevent plastic-shrinkage cracking and excessive heat buildup)
Floor slabs	Curing compound or polythene or other impervious sheeting	As above
Tops of beams and columns	Polythene or other impervious sheeting material	Immediately after finishing process is complete
Concrete columns, beams, walls, etc. in hot dry conditions	Resin-based curing compound	Immediately formwork is removed (some curing compounds may cause debonding of subsequently applied finishes)
	Polythene or other impervious sheeting material	Apply in close contact with surface immediately formwork is removed
	Formwork itself	Leave undisturbed for at least 7 days and longer in cold weather
Formed, permanently exposed concrete sections cast in cold weather	Insulation	As soon as concrete is placed, and maintain for at least 7 days
	Delayed removal of formwork	
Large concrete sections with a minimum dimension exceeding 1 m	Top surface insulation Delayed removal of formwork or replacement of formwork by insulating material	Maintain for at least 7 days and until internal temperature gradient is less than 20°C

Source: Adapted from Birt, J. C., 1984, *Curing Concrete*, Concrete Society (Digest No. 3), London, UK; From Owens, G. (ed.), 2009, *Fulton's Concrete Technology*, 9th edition, Cement & Concrete Institute, Midrand, South Africa, based on Table 12.3, reproduced with permission from the Concrete Institute.

Note: Some manufacturers of resin-based curing compounds recommend re-wetting formed concrete surfaces before application. To avoid thermal shock, the temperature of water used for this purpose must be close to the temperature of the surface of the concrete.

Table 10.3 Guidance on minimum duration of curing[a]

1	2	3	4	5
	Minimum curing period in days[b]			
	Concrete strength development[c] $(f_{cm,3}/f_{cm,28}) = r$			
Surface concrete temperature t (°C)	Rapid $r \geq 0.55$	Medium $r \geq 0.5$	Slow $r \geq 0.45$	Very slow $r \geq 0.25$
≥ 25[d,e]	5	7	10	15
$25 > t \geq 15$	5	7	10	15
$15 > t \geq 5$	7	10	15	21
$t < 5$	10	14	20	30

Source: Taken from SANS 10100-2, 2009, The Structural Use of Concrete, Part 2: Materials and Execution of Work, 3rd edition, South African Bureau of Standards, Pretoria, South Africa, reproduced with the permission of the South African Bureau of Standards (SABS). The South African Bureau of Standards is the owner of the copyright therein and unauthorized use thereof is prohibited. The standard may be obtained at the following: www.store.sabs.co.za

Note: Some curing compounds inhibit the bond of finishes, such as toppings, plasters, or paints, applied to the hardened concrete. The compound used should therefore be suitable for the intended finish.

[a] Based on 70% of 28-day strength.
[b] Linear interpolation between values in the rows is acceptable.
[c] The concrete strength development is the ratio of the mean compressive strength after 3 days $(f_{cm,3})$ to the mean compressive strength after 28 days $(f_{cm,28})$ determined from initial tests or based on known performance of concrete of comparable composition.
[d] Where the ambient relative humidity is above 85%, this value may be reduced at the discretion of the engineer.
[e] Where the ambient relative humidity is below 85%, this value may need to be increased as directed by the engineer.

states, "In general, when the development of a given strength or durability is critical to the performance of the concrete during construction or in service, the minimum duration of curing shall be established on the basis of tests of the required properties performed with the concrete mixture in question. When no such test data are available, the minimum curing period shall be as shown in Table 4" (reproduced here as Table 10.3). The reference to the duration of curing being established on the basis of tests of the required properties is interesting, particularly in regard to durability properties. Unfortunately, there are as yet no standard *in situ* tests to determine whether the required characteristics have indeed been met. With progress in site techniques such as surface resistivity or permeability/sorption, it may be possible in the future to specify an *in situ* cover property value, which ensures that adequate curing has been achieved. Techniques involving removal of small samples from the cover layer, such as the oxygen permeability index (OPI) test or similar (Chapter 6), are able to measure the properties of the surface layer, and thus permit better decisions around type and extent of curing.

Curing compounds are frequently used to aid in concrete curing. These are membrane-forming liquids sprayed onto the surface of the concrete to inhibit evaporation of water, applied as soon as bleeding has stopped and bleed water has evaporated from the surface. They may be convenient means of curing concrete, but not all curing compounds are equally effective and this aspect must be discussed with the supplier of the particular compound being used. Very few curing compounds are effective against rapid early moisture loss. One study evaluated the practical effectiveness of resin-based compounds on highway barrier walls in Cape Town, using site-based tests on the actual cover layer of concrete. Results showed that during cool wet weather, these materials prevented environmental moisture from entering the concrete, thus stifling a possible means of ongoing curing, while in very hot dry weather, their effectiveness was marginal (Krook and Alexander, 1996). These compounds need to be used in strict accordance with the manufacturer's recommendations with due understanding of their likely benefits.

The Australian Standard, AS 3600 "Concrete Structures" (AS 3600, 2009) uses the concept of "average strength at completion of curing," in an attempt to achieve quality of construction; while this is novel, it suffers from a lack of guidance on acceptable curing methods and the fact that strength is measured on samples made under lab conditions rather than the as-built structure. The same comments apply to Table 10.3. This approach is further problematic as it may discourage the use of SCMs such as slag or fly ash that are known to enhance durability properties but have slower strength development.

10.3.3 Cover control and cover measurement

Cover is the distance from the surface of the concrete to the nearest reinforcing bar surface, including links. There are standard specifications for steel reinforcing and prestressing in various countries, and reinforced concrete design codes give guidance on the reinforcement cover needed to protect it from various exposures. For example, BS 8500-1 and EN 1992-1-1 provide cover depth requirements based on exposure classes defined in EN 206-1. BS 8500-1 provides cover depth requirements for different degradation mechanisms while EN 1992-1-1 gives cover depth for corrosion protection only. For other degradation mechanisms, EN 1992-1-1 stresses concrete compositions and proportions (structural class) outlined in Section 6 of EN 206-1. EN 1992-1-1 prescribes cover requirements in terms of nominal cover that includes minimum cover plus any expected deviation. In Canada, CSA A23.1/23.2 provides for minimum cover depth to reinforcing steel based on the life expectancy of the structure, exposure conditions, protective systems, and consequences of corrosion. SANS 10100-2 provides minimum cover depth to reinforcing steel based on the grade (class) of concrete and the member or surface to which cover applies for different conditions of exposure. This minimum cover depth applies for normal-density and low-density concrete. Guidance on minimum cover from EN 1992-1-1, CSA 23.1/23.2, and SANS 10100-2 standards and for various exposure conditions is summarized in Table 10.4.

Bending and fixing of reinforcement are carried out to certain accuracies and tolerances as specified in the relevant standards. Reinforcement cannot be fixed and bent better than the tolerances allowed, particularly when larger bar sizes are used. Even with preassembled reinforcement cages, difficulty is often experienced in accurately aligning the bars. When this is coupled with the allowable tolerances for assembling of formwork, it is quite easy for cover to be compromised, for example, when a reinforcing cage is assembled to its maximum tolerance and the formwork to its minimum tolerance. Even a small reduction in cover can have serious effects on durability and service life, since the relationship between service life and cover is highly nonlinear. This is illustrated in Figure 10.6: hypothetically, halving a nominal cover of 26 mm can result in an 85% reduction in corrosion-free service life (values are illustrative only).

10.3.3.1 Specification of cover and cover measurement

Numerous codes and standards such as ACI 318, AS 3600, CSA A23.1, EN 206-1, and SANS 10100-2 have provisions for minimum cover depth to reinforcing steel in reinforced concrete structures, although some do not have similar direct links between the exposure environments, classes, or categories and the cover depth. For example, ACI 318 has a very broad link between the minimum cover depth to reinforcing steel and the exposure conditions, which are defined in three categories; these are for concrete cast against and permanently exposed to earth, concrete exposed to earth or weather, and concrete not exposed to weather or in contact with ground. On the other hand, AS 3600 introduces subclasses within particular exposure environments. For example, for concrete members in contact with the ground, the subclasses include members protected by a damp-proof membrane, residential footings in nonaggressive soils, other members in nonaggressive soils, and so on. Further, these exposure conditions are related to other parameters such as the initial duration of continuous curing and the characteristic strength of concrete. This trend of incorporating exposure subclasses and other parameters is mostly adopted by other codes such as CSA A23.1, EN 206-1, and SANS 10100-2.

Various reinforced concrete design codes such as EN 1992-1-1, BS 8500-1, and ACI 318 require that the nominal cover is specified on drawings. The nominal cover is obtained by adding an allowance in design for tolerance to the specified minimum value, subject to the quality assurance scheme available on-site. It is important that the specification of cover is clear and unambiguous, and allowance should be made for any surface details such as recesses, treatment such as bush hammering, or where concrete is to be cast against uneven surfaces.

Table 10.4 Guidance on minimum cover for various exposures

Exposure classification/degree of exposure	Reinforced concrete design code: Minimum cover (mm)													
	EN 1992-1-1						CSA 23.1/23.2[a]			SANS 10100-2[b]				
	Structural class						Exposure condition			Grade of concrete (MPa)				
	S1	S2	S3	S4	S5	S6	E*	B*	S*	20	25	30	40	50
X0	10	10	10	10	15	20	—	—	—	—	—	—	—	—
XC1	10	10	10	15	20	25	—	—	—	—	—	—	—	—
XC2/XC3	10	15	20	25	30	35	—	—	—	—	—	—	—	—
XC4	15	20	25	30	35	40	—	—	—	—	—	—	—	—
XD1/XS1	20	25	30	35	40	45	—	—	—	—	—	—	—	—
XD2/XS2	25	30	35	40	45	50	—	—	—	—	—	—	—	—
XD3/XS3	30	35	40	45	50	55	—	—	—	—	—	—	—	—
N[c]							—	30	20					
F-1, F-2, S-1, S-2							75	40	40					
C-XL, C-1, C-3, A-1, A-2, A-3							75	60	60					
Moderate										50	45	40	30	25
Severe:										NA	50	45	40	35
• All exposed surfaces														
• Surfaces on which condensation takes place														
• Surfaces in contact with soil or permanently under running water														
• Cast in situ piles:														
• Wet cast against casing										50	50	50	50	50
• Wet cast against soil										75	75	75	75	75
• Dry cast against soil										75	75	75	75	75

(Continued)

Table 10.4 (Continued) Guidance on minimum cover for various exposures

Exposure classification/degree of exposure	Reinforced concrete design code: Minimum cover (mm)													
	EN 1992-1-1						CSA 23.1/23.2ᵃ			SANS 10100-2ᵇ				
	Structural class						Exposure condition			Grade of concrete (MPa)				
	S1	S2	S3	S4	S5	S6	E*	B*	S*	20	25	30	40	50
Very severe:														
• All exposed surfaces of structures within 30 km from the sea										NA	NA	NA	60	50
• Surfaces in rivers polluted by industries										NA	NA	NA	60	50
• Cast in situ piles, wet cast against casings										NA	NA	NA	80	80
Extreme										NA	NA	NA	65	65

Note: X0, no risk of corrosion or attack; XC1, XC2, XC3, and XC4, corrosion induced by carbonation; XD1, XD2, and XD3, corrosion induced by chlorides other than from sea water; XS1, XS2, and XS3, corrosion induced by chlorides from sea water; C-XL, C-1, and C-3, exposed to chlorides; F-1 and F-2, exposed to freeze/thaw conditions; A-1, A-2, and A-3, exposed to chemical attack; S-1 and S-2, exposed to sulfate attack; N, exposed to neither chlorides nor freeze/thaw conditions; E*, cast against and permanently exposed to earth; B*, beams, girders, columns, and piles; S*, slabs, walls, joists, shells/plates.

ᵃ Greater cover or protective coatings might be required for exposure to industrial chemicals, food processing, and other corrosive materials.

ᵇ The cover values are characteristic minimum values and not more than 5% of cover measurements should fall below these values. In addition, no single cover measurement should fall below 5 mm less than the relevant cover value indicated above.

ᶜ This refers only to concrete that will be continually dry within the conditioned space (i.e., members entirely within the vapor barrier of the building envelope).

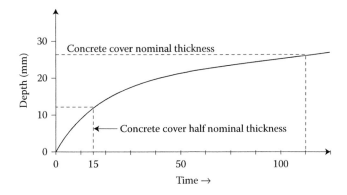

Figure 10.6 Reduction of initiation time of corrosion due to local reduction in concrete cover (values illustrative only). (After Owens, G. (ed.), 2009, *Fulton's Concrete Technology*, 9th edition, Cement & Concrete Institute, Midrand, South Africa. Figure 14.3, reproduced with permission from The Concrete Institute.)

When designing for durability requirements, there is a need to consider the actual minimum cover required to provide the level of protection assumed in the design. It should also be possible to take into account the cover achieved in practice through consideration of an acceptable probability of the cover being less than a certain value.

Concrete cover is generally measured through the use of electromagnetic cover meters that comply with, for example, BS 1881-204. They should be capable of identifying both the location and the depth of reinforcing steel on a scanned area. However, many factors, both device-based and operator-based, can influence the output from such devices, possibly leading to erroneous cover evaluations. It is therefore essential to calibrate such devices against a series of known standard cover depths.

Figure 10.7 shows a reinforcing steel cage with fixed spacers in vertical (wall) formwork, as well as problems associated with insufficient concrete cover.

10.3.3.2 Spacers or spacer blocks

Practice around the use of spacer blocks is highly variable. The reality is that poorly executed spacers represent a short circuit to the steel if they are particularly permeable or a void is created during placing and compaction due to their presence or type. Other problems relate to spacers that can crush, break, or dislodge during heavy handling and poor placing. Spacers that are displaced inside the forms can result in tie wire embedded in the spacer extending effectively to the surface, and then readily corroding. Particular care needs to be taken in obtaining high-quality spacers, and placing and securing them properly during construction.

(a)

(b)

Figure 10.7 Cover to steel in reinforced concrete structures. (a) Spacers in vertical (wall) formwork. (b) Problems associated with insufficient concrete cover.

Adequate impermeable spacers of the correct size must be used and must be firmly fixed to prevent displacement during concreting. Spacer blocks should be made of at least the same grade of concrete used in the structure, with the coarse aggregate sieved out. Alternatively, they can be made from plastic with suitable profiles to allow effective embedment (although some authorities believe plastic spacers represent a short circuit to the steel and forbid their use). If cementitious materials are used for spacers, they must be manufactured well in advance and thoroughly cured in a controlled curing tank prior to use. Figure 10.8 gives examples of (a) commercially available plastic spacers fixed into position and (b) the right and wrong way to install wire ties in a rounded-end, conventional mortar type spacer.

Figure 10.8 (a) Examples of plastic spacers in use and (b) right and wrong installation of wire ties in a mortar spacer. (Courtesy of G. Evans.)

10.4 CONCRETING IN COLD AND HOT CONDITIONS

Concreting in cold and hot conditions presents particular challenges, not only in respect of obtaining required strength, but also in terms of obtaining needed durability of the cover layer. Further useful detail can be found in ACI 306R-10 and ACI 305R-10; however, the following pertinent points should be noted.

10.4.1 Concreting in cold weather

- Concrete should in all circumstances be prevented from freezing in the plastic state or in the very early stages of setting and hardening. If the concrete achieves a "green strength" of about 3–5 MPa, it can withstand a freezing cycle. Bear in mind that at temperatures at or near freezing, the setting and hardening reactions will be considerably retarded and the concrete will take much longer to achieve its properties.

- The minimum concrete temperature as mixed depends on the application and the actual external temperatures. For mass concreting, lower minimum temperatures are acceptable. In general, for air temperatures between −18°C and −1°C, recommended concrete temperatures vary between 7°C and 18°C. The differential between the concrete temperature as placed and the ambient temperature should also be controlled to prevent thermal shock and excessive thermal gradients. Note also that concrete placed at low but nonfreezing temperatures (5–13°C), protected against freezing and given adequate long-term curing, develops a higher ultimate compressive strength.
- Concerning protection and curing at low temperatures, heat loss from the freshly placed concrete should be prevented, and immediate water curing should be avoided. Heat may be retained by insulated forms, covering exposed surfaces with insulating materials, or the erection of covers with internal heating. Formwork and props must be left in place longer than in normal weather.
- Provided the protection mentioned above is given to concrete placed in cold conditions, the interior of the concrete should develop acceptable microstructure and properties. However, as in all cases of concrete placement and the early period, the surface "skin" of concrete is particularly vulnerable in cold conditions. This layer will be the first to experience freezing if it occurs, any heat exotherm generated internally may not diffuse to the outer skin, and cold conditions may also be dry, which further exacerbates the need to protect the outer cover layer. Thus, particular care needs to be taken with protecting the surface of concrete in cold conditions and avoiding loss of moisture, particularly during the early stages of setting and strength development.
- Attention must be given to aspects such as cement type, aggregate protection or heating, lagging of water pipes, heating of mixing water, and batching and mixing equipment during cold weather. Details can be found in ACI 306R-10.
- Concrete should not be placed on a 5°C and falling scale, unless adequate protection against the effects of freezing is applied. Placing of concrete should commence only when ambient temperatures are above 3°C and on a rising scale.

10.4.2 Concreting in hot weather

Ambient temperatures above 32°C are normally taken to be "hot weather." However, any conditions involving temperatures exceeding about 25°C together with low relative humidity and high wind velocity, high solar radiation, or high concrete temperatures can also be treated as hot weather concreting, for which the main problem is rapid loss of water from the concrete and acceleration of early cementing reactions. These will result in reduced concrete quality, and once again, the surface cover zone will suffer the most if proper protection and curing are not provided. The adverse consequences of inadequate protection of concrete during hot weather are likely to be more severe for concrete durability than for concrete strength.

Other aspects to be considered are the following:

- During hot weather, the rate of slump loss and the incidence of plastic-shrinkage cracking increase, setting time decreases, and water requirement for a given consistence increases.
- Early-age strengths tend to be higher but long-term strengths and impermeability are significantly lower. Figure 10.9 shows the effect of elevated concrete temperature on both the early and long-term strength gain. Cements containing extenders such as FA or GGBS tend to be less affected than PC mixes, but require greater attention to protection and curing systems.

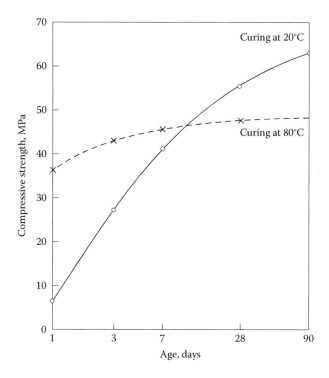

Figure 10.9 Effect of curing temperature on strength of concrete at ages of 1–90 days (for concrete with a w/c ratio of 0.47). (From Murdock, L. J., Brook, K. M., and Dewar, J. D., 1991, *Concrete Materials and Practice*, 6th edition, Edward Arnold, London, UK. With Permission from Hodder Arnold.)

- It is important to control concrete temperatures during hot weather. This can involve some or all of the following: cooling aggregate stockpiles; keeping mixing water as cool as possible; using refrigerated water or even injected liquid nitrogen directly into the concrete mixer or truck mixer to reduce concrete temperatures, or using crushed ice as part of the mixing water; keeping equipment cool; reducing transport times to a minimum; using retarding admixtures; and selecting an appropriate cement incorporating FA or GGBS.

Further useful information can be found in ACI 305R-10.

10.5 QUALITY CONTROL FOR DURABILITY

Chapter 6 indicated that durability of reinforced concrete structures is largely related to the quality and thickness of the cover layer. The required concrete durability must also be established with reference to the structure and the environmental exposure. The desired end is a durable structure—that is, a structure that will endure for the required life in the design environment with acceptable levels of maintenance. The desired properties of the concrete cover layer must be quantified by the owner or designer. Bulk properties such as compressive strength are inadequate for specifying cover quality, and other means must be used such as durability indicators (see Chapter 6). Limits for these indicators are needed in the specification to help ensure durability.

In practical construction, the responsibility for producing a quality structure is shared by the concrete material supplier and the constructor. Therefore, a distinction must be made between material potential and as-built quality of the cover layer (discussed in Chapter 6). The former is the potential of the material to be durable (i.e., what can be achieved), while the latter is the durability exhibited by the structure using the material in service (i.e., what is actually achieved). Durability performance will depend on both these aspects, and they must be allowed for in design and construction. Means need to be put in place for evaluating concrete cover quality at the material production phase and at the construction phase.

As-built durability performance is typically more sensitive to construction processes than as-built strength performance. Except in cases of gross under-compaction, the internal concrete core, which primarily provides strength, is much less affected by most construction processes.

10.5.1 Evaluating material potential quality

To evaluate the material potential, standard specimens (e.g., cubes or cylinders) should be prepared from the concrete supplied, and cured appropriately, which might involve standard curing after demolding. This might be the "standard" 28 days, or a greater (or lesser) period if desired. The specimens should then be tested for the cover properties. The object is to evaluate the representative properties of the concrete under "best scenario" conditions in order to determine if it has the potential to be durable. If concrete is being supplied by a party other than the constructor, for example, by a ready-mix supplier, the supplier should verify that concrete of the desired quality has been provided. Clearly, contractual arrangements in respect of payment for concrete supplied versus concrete placed in the actual structure need to be devised in advance, as well as the responsibility for testing, to avoid subsequent disputes.

As a general rule, concrete in the as-built structure may be of lower quality compared with the same concrete cured under controlled laboratory conditions. To account for this, the characteristic values for the durability properties of the laboratory concrete should represent higher quality on a statistical basis. This will permit some margin between as-delivered quality and as-built quality to allow for any reduction in quality during construction.

10.5.2 Evaluating as-built concrete quality

To evaluate the combined effects of material potential and on-site processing, samples should be taken from the structure (or some appropriate test member that is representative of the actual structure) at an agreed age and tested for the durability properties. This age may be the normal 28 days, but longer if blended cements are used, for which properties take longer to develop. Alternatively, a technology that is available and approved for *in situ* determination of cover properties (such as a Torrent Meter or surface resistivity) can be used. The characteristic values of these properties should be determined, which in general may be somewhat "poorer" than the values achieved for material potential. Other aspects to be considered in this general scheme are the following:

- *Testing frequency for material potential and as-built values*: This will depend on the importance of the job, capacity to carry out testing, and representativeness of the actual construction.
- *Testing locations*: Small areas with obvious physical defects should be avoided. Furthermore, unavoidable spatial variations in concrete quality across a section (e.g., vertical variations in a single pour height of a wall) should be acknowledged and accounted for.
- *Conformity criteria and conformity control*: Conformity control involves the combination of actions and decisions taken in accordance with conformity rules, in order to check the conformity of the concrete with the construction specification. Conformity control is an integral part of production control. Conformity criteria must be agreed in advance and clearly communicated and understood by all

parties. The possibility of incentives for excellent performance and penalties for poor performance can also be considered.

For example, EN 206-1 deals with conformity control in Clause 8, and other standards have similar clauses dealing with this issue, such as CSA A23.1. EN 206-1 covers conformity criteria for compressive strength (individual results and mean results), tensile splitting strength, and other properties of concrete such as consistence, viscosity, passing ability, segregation resistance, and air content. Important points are the following:

- The properties of concrete for conformity control are measured by appropriate tests using standardized procedures. The actual values of the properties of the concrete in the structure may differ from those determined by the tests depending on, for example, dimensions of the structure, placing, compaction, curing, sampling, and environment.
- A sampling and testing plan and conformity criteria should be drawn up in advance.
- Sampling positions for conformity tests should be chosen such that the relevant concrete properties and concrete composition do not change significantly between the place of sampling and the place of delivery. Some specifications identify the point of sampling to be as close as possible to the point of discharge so that the sample represents the concrete being cast into the structure (but then, the as-built properties cannot be evaluated!)
- Tests for production control may be the same as those required for conformity control.
- Conformity or nonconformity is judged against the conformity criteria. Nonconformity may lead to further action at the place of production and on the construction site.
- The relevant standards should be consulted for further details.

The above is necessarily brief, and further guidance can be found in the following references: RILEM TC 189-NEC (2007), Alexander et al. (2008), RILEM TC 230-PSC (2016), and CSA A23.1/A23.2.

10.6 CASE STUDIES OF DURABLE CONSTRUCTION

This section contains two examples from recent large infrastructure projects in which durability was a dominant feature. Although brief, they indicate different ways in which owners and designers have grappled with the problem of actually achieving structures that will be durable for a substantial period of time. Importantly, in these projects, the owners foresaw the durability challenges and were explicit in requiring rational engineering approaches to overcome these challenges.

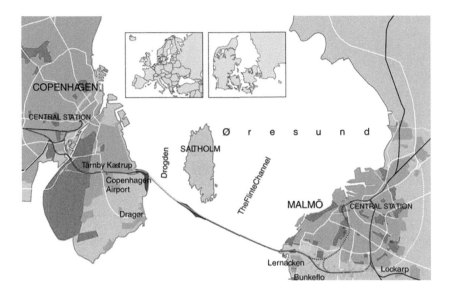

Figure 10.10 Øresund Link between Denmark and Sweden.

10.6.1 Concrete bridge, Øresund Link*

10.6.1.1 Background and design

The 16-km Øresund Link between Copenhagen in Denmark and Malmö in southern Sweden has numerous and substantial structures associated with it, including a tunnel and concrete bridge (Figure 10.10) (Gimsing and Georgakis, 2012; Russel, 2000; Munch-Petersen, 2003). Construction on this prestigious project ran between 1995 and 2000. The bridge of interest is 7.8 km long, making it Europe's longest combined road and railway bridge. It comprises a 490-m cable-stayed span with a navigational height of 57 m, the largest cable-stayed bridge span in the world in terms of length combined with traffic load (at the time of writing). The western approach has 22 spans (4 spans of 120 m each, and 18 spans of 140 m each), and the eastern approach has 27 spans (3 spans of 120 m each, and 24 spans of 140 m each). The bridge superstructure has an upper level with a four-lane motorway, and a lower level with a dual track railway, separated by two parallel Warren-type steel trusses in composite action with the transversely prestressed concrete top flange. A photo of the completed bridge is given in Figure 10.11.

---

* This section is based on a chapter by Braestrup (2016), "Danish strait crossings: Lillebælt, Storebælt, Øresund, Femern Bælt," in *Marine Concrete Structures—Design, Durability, and Performance*, edited by M. G. Alexander, Woodhead Publishing, an imprint of Elsevier, 2016.

Figure 10.11 Completed Øresund Link Bridge. (Courtesy of Henry von Platen/CC, 2013.)

The project was undertaken as a "design and construct" project, with the detailed design done by the contractors based upon conceptual designs and the contract documents. These documents were prepared by the owner, assisted by in-house consultants, and included design and construction requirements, and the functional and durability criteria for the bridge. The design requirements encompassed the design basis and codes and standards (including the Eurocode Project Application Document). The functional criteria were in the form of definition drawings specifying dimensions (e.g., sections, profiles), as well as an illustrative design, prepared by the in-house consultants, showing one way of satisfying the criteria. The durability criteria, included in the construction requirements, specified the design life (120 years), the mix and cover design, and requirements for simulation of early-age cracking and an extensive program of pretesting and full-scale trial casting.

10.6.1.2 Concrete materials, mix designs, and other relevant information

Some typical concrete mix designs and materials are given in Table 10.5. The contractor was permitted to adapt the concrete to the chosen method of execution, subject to the following requirements: minimum total binder content—340 kg/m³; minimum Portland cement content—325 kg/m³ (bridge); maximum PFA content—0% (bridge) and 15% (tunnel) (efficiency factor 0.3); maximum MS (microsilica) content—5.0% (efficiency factor 2.0); and maximum water/binder ratio—0.40 (Type A—see Table 10.5). The specification was developed by a joint Swedish–Danish working party, and the restrictions on the use of fly ash in the bridge structure (but not the tunnel) reflect Swedish reserve with this mineral addition. The bridge structures used concrete type A, from the lower limit of the splash zone (level −3.0 m) to a level 6.0 m above the deck.

Table 10.5 Typical concrete compositions for the Øresund Link

Concrete type	A	A*
PC	375	380
PFA	–	–
MS	12.5	19
Water	145	134
Sand 0–2 mm	647	561
Stone 2–8 mm	288	141
Stone 8–16 mm	439	447
Stone 16–25 mm	466	621
Water/binder ratio	0.36	0.32

Note: A*, concrete mix for concrete road deck. All values (except water/binder ratio) are in kg/m³ (dry mass); amounts of admixtures (air entrainer, retarder, plasticizer, super-plasticizer, and stabilizer) are omitted.

The specified minimum cover to the reinforcement was 30 mm (internal faces in box girders), 50 mm (structures below the splash zone, external faces under waterproofing, internal faces in general), and 75 mm (external faces in general).

Only ordinary black steel reinforcement was used, in addition to pre-stressing steel. For prestressing steel, the concrete cover was generally speci-fied as 5 mm greater than the values mentioned above.

10.6.1.3 Construction and quality management

Computer simulations were used to predict early-age temperatures and stresses in the concrete sections. No early-age cracking was accepted for parts of the bridge structure in the splash zone. For other parts, the maxi-mum allowable early-age crack width was 0.2 mm.

Significantly, the only component of the bridge constructed *in situ* was the bridge towers, with the rest of the elements being prefabricated to help ensure quality of construction. For construction of the 204 m high tower legs, a floating concrete production plant was anchored alongside the bridge, with the use of climbing formwork in lifts of 4 m, the concrete being lifted by buckets, with a cycle time of 5–8 days in two shifts. On the western pylon, some cracking was observed on early pours, and on parts of the northern leg, the concrete did not meet required standards. The cracks were repaired, but two lifts on the northern leg were demolished and recast.

During all phases of production, concreting, and curing, strong empha-sis was placed on pretesting and production testing, and on the contrac-tor's quality management systems, including inspection plans and quality documentation. Pretesting comprised standard testing of constituents and trial mixes (including air content and salt scaling by the Swedish Borås method), as well as full-scale casting to prove that the selected production method matched the intended concrete mix. Production testing included air content and salt scaling. Furthermore, the installation of corrosion monitors was specified.

The concrete caissons, pier shafts, and troughs for the railway deck of the approach bridges were constructed using onshore prefabrication, allowing higher quality control and timeous production of these elements. Approach span steel trusses, including the concrete road deck, were prefabricated in Cadiz (southern Spain), using concrete type A* (see Table 10.5), and transported to the site on sea barges. The concrete strength for the road deck was 50 MPa, with air entrainment for frost resistance. The requirement was a minimum of 3% air in hardened concrete unless the specified pretesting had shown that higher air contents were necessary in order to comply with the requirements for frost scaling (max. 0.5 kg/m^2 after 56 cycles; salt scaling testing according to Swedish Standard SS 137 244 (the Borås method, 2005)).

Figure 10.12 shows a photograph of a road span being hoisted onto the bridge.

The quality management approach required the contractors to approve their own work. They executed quality system audits on all aspects of the work, including subcontractors and suppliers, to check whether the contractor's actual execution of the various activities complied with their quality plans. The owner's staff also conducted regular audits on predefined areas of each contractor's activities. Ultimately, the two audit programs

Figure 10.12 Øresund Bridge, approach span erection using the heavy-lift vessel "Svanen."

were coordinated to avoid unnecessary duplication; interestingly, as cooperation evolved, the contractors invited the owner and his or her consultants to attend their audits as observers.

10.6.1.4 Inspection and maintenance

A detailed inspection plan for the concrete structures was prepared, specifying inspection periods, test procedures, monitoring of the structure, and recording of observations. The inspection plan also included actions to be taken after inspection, such as modified inspection frequencies, initiation of cathodic protection, and repair or replacement of elements.

In brief, inspection was needed to confirm that drainage of water occurred from all concrete surfaces for the roadway bridge deck and the railway containment troughs, which are exposed to freezing and some chlorides from deicing salts. The edge beams in particular are highly exposed, and therefore subjected to detailed inspection. Minor repairs of edge beams may be envisaged in the future, with replacement being considered as an alternative to larger repairs.

The bridge piers and pylons are exposed to chlorides, to freezing up to some distance above the bridge deck, and also to deicing salts at the deck level. Consequently, exposed parts of pylons have been prepared for cathodic protection and equipped with corrosion monitors for timely activation of the cathodic protection. If needed, cores and samples may also be taken to verify the chloride transportation model.

The bridge abutments and viaduct are in some areas exposed to chlorides, freezing, and deicing salts. They were handled in a similar way to the bridge piers and pylons. The viaduct beam elements are less exposed and possible repairs may be limited to edge beams only.

10.6.1.5 Summary

The Øresund Link Bridge was conceived, designed, and constructed with clear durability considerations in mind. Considering best practice at that time, concrete materials and construction methods were chosen and a specification put in place to ensure maximum protection against the main causes of deterioration to the bridge structure and its elements: chloride ingress and freeze–thaw. Use of a supplementary cementitious material in the form of microsilica (silica fume) was considered essential for long-term durability. Although the structures were built using essentially a prescriptive approach, much emphasis was placed on pretesting and production testing and on the contractors' quality management systems, including inspection plans and quality documentation. A regular inspection and maintenance program is in place to detect problems timeously and limit possible repair costs.

10.6.2 Construction of the new Panama Canal*

10.6.2.1 Introduction: background, structures, and environment

The original Panama Canal was built 100 years ago, comprising a canal from the Pacific side to Lake Gatun (26 m above sea level) in the interior, with a shorter canal from Lake Gatun to the Gulf of Mexico, and several locks.

For decades, there has been recognition of the need to enlarge the width of the canal to take wider and heavier ships. The Panama government finally decided to undertake the upgrade, and during 2006, a tender for the construction of two lanes for the new locks was issued. The construction project was awarded to a consortium Grupo Unidos por el Canal (GUPC), with completion scheduled for 2016.

The new construction, from a concrete perspective, involved casting about 4.7 million cubic meters of concrete representing, at the time, the world's largest single concrete production site. Concrete plants were installed on each site, with nominal capacities of 540 m^3/hour. The Pacific site is shown in Figure 10.13.

The coarse and fine aggregates comprised basalt quarried from the site quarry, and processed in crushing plants. The aggregate was not as sound as expected, producing a large amount of "superfine" material. Later, this superfine fraction, after verifying its pozzolanic activity, was added to the concrete to improve chloride resistance.

The exposure environment for the concrete structures differed depending on the locality: the lock nearer the sea (i.e., the lower chamber) has salinities of up to 34 g/L, while the interior sections adjacent to Lake Gatun, which contains fresh water (i.e., the upper chamber), had salinities as low as 0.6 g/L. These values can be compared to typical seawater salinities of around 35 g/L, although this varies widely in different oceans.

10.6.2.2 Concrete design philosophy and basis

Concrete for the new canal structures required to be reinforced as an antiseismic precaution. Two concrete types were specified: marine and standard concrete. The marine concrete, placed in the external exposed surfaces of the elements, had more stringent requirements. The concrete had to be cast on site and be pumpable. The maximum w/b ratio was 0.4.

The specification of the Panama Canal Authority required a 100-year service life for the marine concrete in all members, and also a maximum 1000-coulomb criterion according to ASTM C1202. However, no details were given on what the 100-year service life meant, or how to demonstrate

---

* This section is based on a chapter by Andrade et al. (2016), "Concrete durability of the new Panama Canal: Background and aspects of testing," in *Marine Concrete Structures— Design, Durability, and Performance*, edited by M. G. Alexander, Woodhead Publishing, an imprint of Elsevier, 2016.

Figure 10.13 View of the Pacific concrete plant for the new Panama Canal, with the silos and the refrigeration plant.

it, although calculations regarding chloride penetration were required. Minimum cover depths of 75 mm were specified, with requirements for temperature control of not higher than 70°C in the center of the elements to avoid thermal cracking. This necessitated the use of ice in the mixes during casting, and temperatures were checked using thermocouples in the fresh and hardening concrete, together with theoretical heat development studies at element size. A curing time of 14 days was also prescribed.

10.6.2.3 Durability aspects and critique of specification

The specification requirements for assuring the 100-year service life were the following:

For all concretes irrespective of exposure class:

- A maximum of 1000 coulombs in ASTM C1202 (rapid chloride permeability test [RCPT]). However, there was no indication of the age at which this was to be met.
- Minimum of 100 years to corrosion initiation.
- Low shrinkage cracking (a maximum temperature of 70°C in the cast concrete in the early stages of hydration, and less than 20°C difference in the interior of the concrete).

Further requirements of the specification were a maximum w/b ratio of 0.40, as stated before, and concrete shrinkage testing according to ASTM C157.

Concerning the minimum 100-year corrosion-free life, a service life prediction model was suggested, which was accepted by the consortium building the canal (Andrade and Whiting, 1996; Andrade, 2004; Andrade and Tavares, 2012; Andrade et al., 2014). In general, the specification had shortcomings and limitations, leading to the following critique:

1. The deterioration mechanism for which there was a coulomb limitation was not defined. This criterion was linked with "permeability" and "durability" requirements.
   a. No tests directly measuring the diffusion coefficient of chlorides were prescribed.
   b. No age was specified for the conformity requirement. The "aging" of the concrete was not taken into account.
2. The characteristics or basis for the 100-year calculation of service life were not specified.
3. No link was established between the coulomb limitation and the 100-year service life requirement.
4. There was no mention of a test relating to the calculation method.
5. No tolerance for the 1000-coulomb was specified, and it was not explicit whether this was a mean-coulomb or characteristic value.
6. No tolerance or accepted deviation on the lifetime was given.
7. The same concrete quality was required for all water salinities in spite of the fact that they were very different. Two of the newly constructed chambers contain waters with chlorides insufficient to induce depassivation.

10.6.2.4 Construction, testing, and verification

Construction work on the new canal started in January 2011, with a laboratory in charge of verifying compliance with the 1000-coulomb electrical charge and calculating the 100-year service life. For the latter requirement, the consortium engaged the services of IETcc,* who suggested

---

* El Instituto de Ciencias de la Construcción Eduardo Torroja (IETcc), Madrid, Spain.

using alternative approaches to ensure compliance with the specifications. These entailed use of different primary materials and batching processes to manufacture a series of alternative concrete mixes; testing natural chloride diffusion (ponding tests); continuous monitoring by measuring electrical resistivity over time; and the use of LIFEPRED, a numerical model developed at IETcc for calculating service life based on Fick's law (Andrade et al., 2011).

An experimental program was developed between GUPC and IETcc, involving testing of concrete mixes from the construction sites for RCPT value, resistivity, natural chloride diffusion (ponding), water penetrable porosity, and mercury intrusion porosimetry (MIP). Ultimately, four concrete mixes from the Pacific and Atlantic sites, out of more than 16 mixes studied, were selected for final testing. A "natural pozzolan" supplied by the cement producer and silica fume was also used in the mixes. Interestingly, the "superfine" fraction of the basaltic crusher sand was found to have pozzolanic properties and was used in the mixes. Final w/b ratios were of the order of 0.3. It was also decided to study concrete from the old canal by way of drilled cores, for chloride profiles and electrical resistivity, MIP, and DTA/TG.

In summary, the results showed that it was possible to replace measuring electrical charge (coulombs) in the RCPT test, with measurements of resistivity. The natural diffusion tests permitted the accurate calibration of the Fick's Law Model used in the LIFEPRED Program. Mixes that would achieve the 100-year corrosion-free service life requirement according to the modeling could then be more confidently selected and used in construction. Of course, it was realized that there are no 100-year data available to verify the model, but the requirements of the client were fulfilled. The various critiques of the specification were also a valuable learning experience.

REFERENCES

ACI 117-10, 2010, *Specification for Tolerances for Concrete Construction and Materials and Commentary*, American Concrete Institute, Farmington Hills, MI.

ACI 304 R-00, 2009, *Guide for Measuring, Mixing, Transporting, and Placing Concrete*, American Concrete Institute, Farmington Hills, MI.

ACI 305R-99, 2010, *Guide to Hot Weather Concreting*, American Concrete Institute, Detroit, MI.

ACI 306R-88, 2010, *Guide to Cold Weather Concreting*, American Concrete Institute, Farmington Hills, MI.

ACI 318-14, 2014, *Building Code Requirements for Structural Concrete*, American Concrete Institute, Farmington Hills, MI.

Alexander, M. G., Ballim, Y., and Stanish, K., 2008, A framework for use of durability indexes in performance-based design and specifications for reinforced concrete structures, *Materials and Structures*, 41(5), 921–936.

Andrade, C., 2004, Calculation of initiation and propagation periods of service-life of reinforcements by using the electrical resistivity, *International Symposium on Advances in Concrete through Science and Engineering*, Evanston, March 22–24.

Andrade, C., Castellote, M., and d'Andrea, R., 2011, Measurement of ageing effect of chloride diffusion coefficients in cementitious matrices, *Journal of Nuclear Materials*, 412, 209–216.

Andrade, C., d'Andrea, R., and Rebolledo, N., 2014, Chloride ion penetration in concrete: The reaction factor in the electrical resistivity model, *Cement and Concrete Composites*, 47, 41–46.

Andrade, C., Rebolledo, N., Tavares, F., Pérez, R., and Baz, M., 2016, Concrete durability of the new Panama Canal: Background and aspects of testing, in *Design and Durability of Marine Concrete Structures*, M. G. Alexander (ed.), Elsevier, Oxford.

Andrade, C. and Tavares, F., 2012, *LIFEPRED—Service Life Prediction Program*, Ingenieria de Seguridad Y Durabilidad S.L., Madrid, Spain.

Andrade, C. and Whiting, D. A., 1996, Comparison of chloride ion diffusion coefficients derived from concentration gradients and non-steady state accelerated ionic migration, *Materials and Structures*, 29, 476–484.

AS 3600, 2009, *Concrete Structures*, Standards Australia, Sydney, New South Wales.

ASTM C1202-12, 2012, *Standard Test Method for Electrical Indication of Concrete's Ability to Resist Chloride Ion Penetration*, Vol. 04.02, ASTM International, West Conshohocken, PA.

ASTM C157/C157M-08, 2008, *Standard Test Method for Length Change of Hardened Hydraulic-Cement Mortar and Concrete*, ASTM International, West Conshohocken, PA.

ASTM C94M-14b, 2014, *Specification for Ready-Mixed Concrete*, American Society for Testing and Materials, West Conshohocken, PA.

Birt, J. C., 1984, *Curing Concrete*, Concrete Society (Digest No. 3), London, UK.

Braestrup, M., 2016, Danish strait crossings: Lillebælt, Storebælt, Øresund, Femern Bælt, in *Design and Durability of Marine Concrete Structures*, M. G. Alexander (ed.), Elsevier, Oxford.

BS 1881-204, 1988, *Testing Concrete. Recommendations on the Use of Electromagnetic Covermeters*, British Standards Institution, London, UK.

BS 8500-1, 2006, *Concrete—Complementary British Standard to BS EN 206-1— Part 1: Method of Specifying and Guidance for the Specifier*, British Standards Institution, London, UK.

BS EN 206-1:2013, *Concrete—Specification, Performance, Production and Conformity, Part 1*, British Standards Institution, London, UK.

CSA A23.1-09/A23.2-09, 2014, *Concrete Materials and Methods of Concrete Construction/Test Methods and Standard Practices for Concrete*, Canadian Standards Association, Toronto, Canada.

EN 1992-1-1, 1994, *Eurocode 2: Design of Concrete Structures—Part 1-1: General Rules and Rules for Buildings*, Euronorm, Brussels.

EN ISO 9001, 2008, *Quality Management Systems—Requirements*, Technical Committee ISO/TC 176, Geneva, Switzerland.

Gaynor, R. D., Meininger, R. C., and Khan, T. S., 1985, *Effect of Temperature and Delivery Time on Concrete Proportions*, NRMCA Pub. No. 171, National Ready Mixed Concrete Association, Silver Spring, MD, 21pp.

Gimsing, N. J. and Georgakis, C. T., 2012, *Cable Supported Bridges: Concept and Design*, 3rd edition, John Wiley & Sons, New York, NY.

Kosmatka, S. H., Kerkhoff, B., and Panarese, W. C., 2003, *Design and Control of Concrete Mixtures, EB001*, 14th edition, Portland Cement Association, Skokie, IL.

Krook, A. and Alexander, M. G., 1996, An investigation of concrete curing practice in the Cape Town area, *SAICE Journal*, 38(1), 1–4.

Munch-Petersen, C. (ed.), 2003, *Danish Civil and Structural Engineering*, Danish Society for Structural Science and Engineering, Copenhagen, Denmark.

Murdock, L. J., Brook, K. M., and Dewar, J. D., 1991, *Concrete Materials and Practice*, 6th edition, Edward Arnold, London, UK.

Owens, G. (ed.), 2009, *Fulton's Concrete Technology*, 9th edition, Cement & Concrete Institute, Midrand, South Africa.

RILEM TC-189 NEC, 2007, Non-destructive evaluation of the penetrability and thickness of the concrete cover, in *State of the Art Report*, R. Torrent and L. Fernandez (eds.), Springer, Dordrecht.

RILEM TC-230 PSC NEC, 2016, Performance-based specification and control of concrete durability, in *State of the Art Report*, H. Beushausen and L. Fernandez (eds.), Springer, Dordrecht.

Russel, H., 2000, *Partnership Pays: Project Management the Øresund Way*, Route One Publishing Ltd, London, UK.

SANS 10100-2, 2009, *Code of Practice for the Structural Use of Concrete, Part 2: Materials and Execution of Work*, South African Bureau of Standards, Pretoria, South Africa.

SANS 878, 2004, *Ready-Mixed Concrete*, 4.1 edition, South African Bureau of Standards, Pretoria, South Africa.

Swedish Standard SS 137 244 (the Borås method), 2005, Concrete testing—Hardened concrete—Scaling and freezing, Swedish Standards Institute, Stockholm, Sweden.

FURTHER READING

Choo, B. S., 2003, Reinforced and pre-stressed concrete, in *Advanced Concrete Technology: Processes*, J. Newman and B. S. Choo (eds.), Butterworth Heinemann, Oxford, UK.

Concrete Institute of Australia, 2007, *Reinforcement Detailing Handbook for Reinforced and Prestressed Concrete*, 3rd edition, Concrete Institute of Australia, Sydney, New South Wales.

Concrete Society, 1999. Towards rationalising reinforcement for concrete structures, Technical Report no. 53, Concrete Society, Berkshire, UK.

Illingworth, J. R., 1972, *Movement and Distribution of Concrete*, McGraw-Hill, London, UK.

King, E. S. and Dakin, J. M., 2001, Specifying, detailing and achieving cover to reinforcement, CIRIA Report C568, Construction Industry Research and Information Association, Berkshire, UK.

Index